NOVAE AND RELATED STARS

ASTROPHYSICS AND
SPACE SCIENCE LIBRARY

A SERIES OF BOOKS ON THE RECENT DEVELOPMENTS

OF SPACE SCIENCE AND OF GENERAL GEOPHYSICS AND ASTROPHYSICS

PUBLISHED IN CONNECTION WITH THE JOURNAL

SPACE SCIENCE REVIEWS

VOLUME 65
PROCEEDINGS

NOVAE AND
RELATED STARS

PROCEEDINGS OF AN INTERNATIONAL CONFERENCE
HELD BY THE INSTITUT D'ASTROPHYSIQUE,
PARIS, FRANCE, 7 TO 9 SEPTEMBER 1976

Edited by

M. FRIEDJUNG

Institut d'Astrophysique, Paris, France

Sponsored by
THE CENTRE NATIONAL DE LA RECHERCHE SCIENTIFIQUE

D. REIDEL PUBLISHING COMPANY
DORDRECHT-HOLLAND/BOSTON-U.S.A.

ISBN-13: 978-94-010-1219-5   e-ISBN-13: 978-94-010-1217-1
DOI: 10.1007/978-94-010-1217-1

Published by D. Reidel Publishing Company,
P.O. Box 17, Dordrecht, Holland

Sold and distributed in the U.S.A., Canada, and Mexico
by D. Reidel Publishing Company, Inc.
Lincoln Building, 160 Old Derby Street, Hingham,
Mass. 02043, U.S.A.

TABLE OF CONTENTS

* Speaker who presented the paper.

_____
* Speaker who presented the paper.

* Speaker who presented the paper.
** Read by K. Nariai.

---

* Speaker who presented the paper.
** Read by K. Nariai.

_____
* Speaker who presented the paper.

---

* Speaker who presented the paper.

PREFACE

Michael Friedjung

Though known since antiquity, novae are still poorly understood and present many problems. There has tended to be a
lack of communication between theoreticians and observers and
between different schools of thought, in spite of the advances
of recent years in certain directions (observations of ordinary
novae at minimum and of dwarf novae, theory of the causes of
the explosion, etc...). The meeting whose proceedings are
contained in this volume was organized to stimulate a confron-
tation between the different ideas and results. The subject
has changed a lot since 1963, when the previous international
meeting was held.
     There were 61 participants at the conference from 17
countries, so very many groups doing research in the field of
novae were represented. The reader will see that the subject
has become more physical (we know for instance that the binary
nature of novae is essential) but much work remains to be done.
There is still a large gulf between theory and observation.
May he find here many new ideas for future research!
     I would like to thank the French Centre National de la
Recherche Scientifique for providing financial help. I must
also thank the other members of the scientific committee (Profs
and Drs Bath, Mustel, Payne-Gaposchkin, Sparks and Warner) and
of the local committee (Audouze, Mrs Steinberg, Vauclair).
This meeting would not have been successful without the extremely
efficient administrative help of Mrs Steinberg. I must thank
various technicians of the Institut d'Astrophysique who gave
practical help during the meeting, and especially Mrs Delin who
had to do a lot of typing, and who was persecuted by me when
typing manuscripts for this volume.

LIST OF PARTICIPANTS

Andrews, P.J.	Royal Greenwich Obs., England
Andrillat, Y.	Obs. Haute Provence, France
Audouze, J.	Obs. Meudon, La. René Bernas, Orsay, France
Bath, G.T.	Dept. Astrophys., Oxford, England
Bosma, P.B.	Free University, Amsterdam, Netherlands
Brecher, K.F.	Dept. of Physics, Mass. Inst. of Tech., U.S.A.
Ciatti, F.	Obs. Asiago, Italy
Collin, S.	Obs. Meudon, France
Cowley, A.	Dom. Astron. Obs., Canada
Crivellari, L.	Obs. Trieste, Italy
Drechsel, H.	Dr. Remeis Sternwarte, Bamberg, W. Germany
Duerbeck, H.W.	Astron. Inst., Bonn University, W. Germany
Edwards, P.J.	Phys. Dept. Univ. Otago, New Zealand
Eggleton, P.P.	Astron. Inst. Cambridge, England
Faulkner, J.	Lick Obs., U.S.A.
Feast, M.W.	SAAO, Cape Province, South Africa
Finzi, A.	Technicon, Haifa, Israel
Friedjung, M.	Inst. Astrophys. Paris, France
Gaposchkin, S.	Center for Astrophys., Cambridge, U.S.A.
Grygar, J.	Astron. Inst. Ondrejov, Czechoslovakia
Houziaux, L.	Dept. Astrophys. Univ. Mons, Belgium
Hutchings, J.B.	Dom. Astron. Obs., Canada
Ilovaisky, S.	Obs. Meudon, France
Kemp, J.C.	Univ. Oregon, U.S.A.
Kraft, R.P.	Lick Obs., U.S.A.
Kupo, I.	Tel-Aviv Univ., Tel-Aviv, Israel
Larsson-Leander, G.	Lund Obs., Sweden
Leibovitz, E.	Tel-Aviv Univ., Tel-Aviv, Israel
Lin, D.N.C.	Inst. Astron. Cambridge, England

MacDonald, J.   Inst. Astron. Cambridge, England
Mumford, G.S.   Tufts Univ., Mass., U.S.A.
Muratorio, G.   Obs. Marseille, France
Mustel, E.   Academy of Science, U.S.S.R.
Nariai, K.   Astron. Obs. Tokyo, Japan
Nave, M.F.F.   Dept. Theoretical Phys. Oxford, England
Pacheco, J.   Obs. Nice, Nice, France
Payne-Gaposchkin, C.   Center for Astrophys., Cambridge, U.S.A.
Prialnik, D.   Tel-Aviv Univ., Tel-Aviv, Israel
Rafanelli, P.   Astron. Inst. Padova, Italy
Reiss, S.   Astron. Inst. Munster, W. Germany
Renson, P.   Inst. Astrophys., Liège, Belgium
Rocca, B.   Lab. Bernas, Orsay, France
Rosino, L.   Astron. Obs. Asiago, Italy
Ruggles, C.L.N.   Dept. Astrophys. Univ. Oxford, England
Scuflaire, R.   Inst. Astrophys., Liège, Belgium
Seitter, W.   Astron. Inst. Munster, W. Germany
Selvelli, P.   Obs. Trieste, Italy
Shara, M.   Tel-Aviv Univ., Dept. Physics, Tel-Aviv,
                        Israel
Shaviv, G.   Tel-Aviv Univ., Dept. Physics, Tel-Aviv,
                        Israel
Smak, J.   Inst. Astron. Warsaw, Poland
Sparks, W.M.   NASA, Greenbelt, U.S.A.
Steiner, J.F.   Sao Paulo, Brasil
Szumiejko, E.   Astron. Obs. Wroclaw, Poland
Thomas, H.T.   Max Planck Inst., Munich, W. Germany
Tylenda, R.   Inst. Astron. Torun, Poland
Vauclair, G.   Obs. Meudon, France
Viotti, R.   Space Lab., Frascati, Italy
Vittone, A.   Astron. Obs., Asiago, Italy
Warner, B.   Dept. Astron. Univ. Cape Town, South
                        Africa
Wegener, H.   Astron. Inst. Munster, W. Germany
Wu, Chi Chao   Dept. Space Research, Groningen, Netherlands
Wyckoff, S.   Royal Greenwich Obs., England

PART I

<u>NOVAE, DWARF NOVAE AND SIMILAR OBJECTS AT</u>
<u>MINIMUM LIGHT</u>

PAST AND FUTURE NOVAE

Cecilia Payne-Gaposchkin

Center for Astrophysics
Cambridge, Massachusetts

A CRITICAL LIST OF NOVAE

The standard catalogue of variable stars (Kukarkin et al.,
1969, 1971, 1974, 1976) recognizes several categories of novae:
supernovae (SN), classical novae (N), subdivided where the observations warrant into fast novae (Na), slow novae (Nb), very
slow novae (Nc), recurrent novae (Nr), nova-like variables (Nℓ),
and dwarf novae, including U Geminorum stars (UG) and Z Camelopardalis stars (Z Cam). We exclude supernovae from the present
discussion, and summarize the content and interrelationships of
the other classes.

The (classical) novae are defined as "hot dwarf stars with
spontaneous increase of brightness with amplitude from 7 to 16
magnitudes." A critical list, compiled on the basis of spectroscopic and photometric data, includes 170 classical and 7 recurrent novae. Only one third of them, 55 in all, have been
observed at minimum light. The distribution of the published
ranges is shown in Table 1.

The table displays a large spread in range. The second
column is based on the discussion by Robinson (1975) of pre-eruption light curves. Several stars exceed the limits defined above:
FS Sct and V 2572 Sgr have ranges less than 7 magnitudes, and
those of V 1500 Cyg (over 18 magnitudes) and CP Puppis (over 16.5
magnitudes) are lower limits. The maximum frequency of range observed is between 11 and 12 magnitudes for the classical novae;
the recurrent novae have ranges less than 9 magnitudes. However,
the data are biased in favor of small ranges, for two thirds of
the novae have not been observed at minimum. It is tempting to

M. Friedjung (ed.), Novae and Related Stars, 3-32. All Rights Reserved.
Copyright © 1977 by D. Reidel Publishing Company, Dordrecht, Holland.

Table 1

Distribution of Published Ranges

Magnitude	Classical novae	Robinson	Recurrent novae
6 to 7	2	2	0
7 to 8	2	2	2
8 to 9	6	2	4
9 to 10	8	6	0
10 to 11	11	2	0
11 to 12	12	8	0
12 to 13	4	3	0
13 to 14	6	2	0
14 to 15	3	1	0
15 to 16	1	0	0

surmise that large range goes with rapid development; however the ranges of the fast novae V 603 Aql ($13^m.9$) and GK Per ($13^m.8$) are comparable with that of the slow nova DQ Her ($13^m.7$).

Some novae with observed ranges less than 9 magnitudes may in fact be recurrent, with cycles short enough to be detected within a reasonable time. We recall that U Sco (1866) was not found to be recurrent until Harvard photographs were examined in 1940, and outbursts in 1906 and 1936 were noted by Thomas (1940). Well-determined small ranges are known for HR Lyr, IM Nor, V 841 Oph and FS Sct, and these stars would repay further study. The maximal magnitudes of HR Del, FM Sgr, V 441 Sgr, V 1016 Sgr, V 1944 Sgr, V 3890 Sgr and DZ Ser are uncertain, and the ranges may well have been greater than those recorded. We recall that RR Pic was a third magnitude star three months before it was discovered visually (although in this case the character of the first observed spectra makes it seem probable that the star had an exceptionally long pre-maximum rise, and that the actual maximum was in fact recorded). A further caveat on published ranges should be noted: until standards became available in the thirties of the present century, magnitudes fainter than 13 were extrapolations, and little more than surmises. Re-examination of the magnitudes of early novae on the original plates would modify some of the results, especially for stars in the southern hemisphere.

For reasons of space, the complete critical catalogue will not be reproduced here. The "Census of Novae" published by the writer (1957) will be adopted, with the deletions listed in Table 2. Table 3 gives a list of novae that should be added to the census, bringing it up to July 1976.

Table 2

Name	New Class		Name	New Class
V 607 Aql	Mira		UW Per	UGem
V 407 Cyg	Mira?		V 939 Sgr	Mira

Name	New Class
V 1017 Sgr	Z And
SW Vul	U Gem

DISTRIBUTION OF GALACTIC NOVAE

The galactic distribution of the accepted novae is illustrated in Figures 1 and 2. The known relation between speed of development and maximal luminosity makes it possible (with plausible

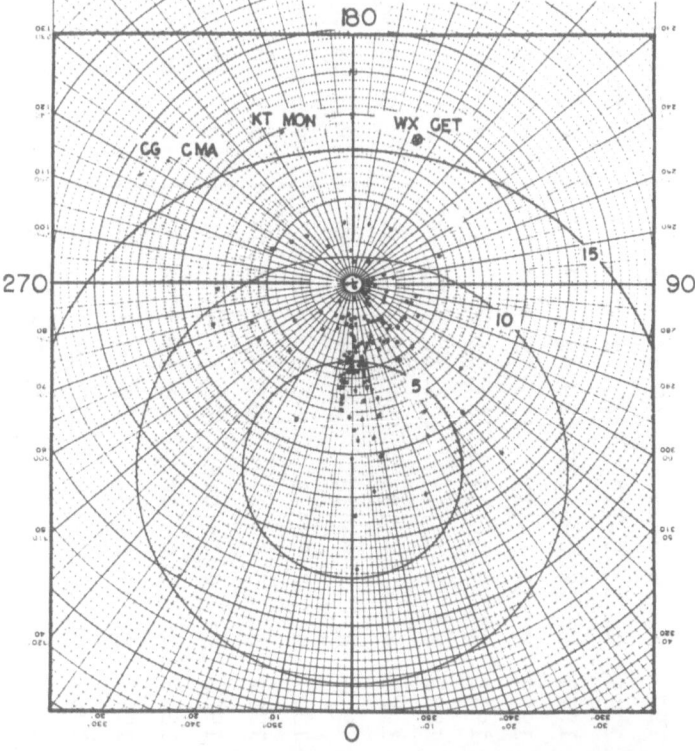

Figure 1. Positions of novae projected on the galactic plane. Galactic longitudes are indicated on the markin and a cross marks the position of the sun. Circles with radii of 5, 10 and 15 kiloparsecs are drawn around the galactic center, adopted as 9 kiloparsecs from the sun. See text for discussion of 3 indicated stars.

Table 3

Additional Novae

Name	Range	l	b	Name	Range	l	b
V 1229 Aql 1970	N 6.5-19	40.54	- 5.44	V 1905 Sgr 1932	Na 9.1-[16.5	8.31	- 7.63
V 1301 Aql 1975	Na 11.5-[19	40.22	- 3.68	V 1944 Sgr 1960	N 7:- 13	3.02	- 1.88
V 351 Car 1970	N? 12.2- 15.2	293.47	- 11.55	V 2415 Sgr 1951	N 13:-[18	0.30	- 1.74
BC Cas 1929	N 12 -[17	115.55	- 1.71	V 2446 Sgr 1953	N? 17.4-[19.3	1.44	- 2.65
IV Cep 1971	Nb 7.5- 17.5	99.61	- 1.64	V 2506 Sgr 1946	N ..	3.10	- 4.64
RR Cha 1953	Na 7.1-[15	304.16	- 19.54	V 2572 Sgr 1969	Na 6.5-[14	1.51	- 10.41
V 655 CrA 1967	N 8:-17:	356.86	- 11.07	V 3465 Sgr 1969	Nb 8:-18	14.52	- 5.09
V 1330 Cyg 1970	N 9-[18	78.38	- 5.49	V 3888 Sgr 1974	N 8:-[13	9.08	+ 4.66
V 1500 Cyg 1975	Na 2.22-[21	89.82	- 0.07	V 3889 Sgr 1975	Na 8.3-20:	1.92	- 2.11
HR Del 1967	Nb 3.5?-12?	63.43	- 13.97	V 3890 Sgr 1962	Na 8.4-17.2	9.21	- 6.44
V 446 Her 1960	Na 3.0-18.8	45.41	+ 4.71	V 744 Sco 1935	N? 13.3-[15.6	358.88	- 2.59
V 533 Her 1963	Na 3.0:-14.9	69.19	+ 24.27	V 745 Sco 1937	N? 11.2-[14.9	357.36	- 3.98
V 616 Mon 1975	Nr 11.3-20	209.96	- 6.54	V 825 Sco 1963	N 8-[13	356.52	- 3.17
V 972 Oph 1957	Nb 8.0-[16.5	359.39	+ 2.43	FV Sct 1960	N 7:-21	19.52	- 2.18
V 1012 Oph 1961	N 14-[21	4.25	+ 3.68	GL Sct 1954	N 13.6-[16.8	26.57	- 1.68
V 1548 Oph 1959	N? 13.7-[20	6.34	+ 13.46	V 366 Sct 1961	N 15.4-[17	19.52	- 0.86
V 2024 Oph 1967	N 9.5:-[18	2.99	+ 2.69	V 368 Sct 1970	Na 7.0-18.6	24.67	- 2.63

Table 3 (cont.)

Name		Range	l	b	Remark
.. Oph 1938	N				
V 400 Per 1974	Nb	8-19.5	145.66	- 9.64	
HS Pup 1963	Nb	8.0-[20	247.76	- 2.11	
HZ Pup 1963	Nb	7.7-18.5	246.18	+ 1.38	
V 1310 Sgr 1935	N?	11.1-15.5	4.14	- 9.97	
V 1431 Sgr 1945	N	17.2-[19.2	1.04	- 3.88	
V 1572 Sgr 1955	N	11-[16	359.82	- 5.08	

Name		Range	l	b
V 373 Sct 1975	Na	6.0-18.5	26.50	- 4.40
DZ Ser 1960	N	8:-16.9	17.62	+ 6.30
FH Ser 1970	N	4.5-16.0	32.91	+ 5.78
RW UMi 1956	Na	6-[21	109.64	+ 33.16
LU Vul 1968	Na	9.2-[21	64.26	+ 2.02
LV Vul 1968	Na	5.7-16.9	63.30	+ 0.85

Remark to Table 3

In Messier 14: Sawyer Hogg (1972).

assumptions about interstellar absorption) to deduce distances
for novae that have been adequately observed, and to obtain less
accurate information for others by adopting a mean luminosity at
maximum. The distribution projected on the galactic plane
(Figure 1) shows a tendency to concentrate in the direction of
the galactic center.

Three stars should be noted in the diagram: KT Mon, whose
spectrum observed by Vyssotsky, and whose light curve obtained
by Gaposchkin (1954) place it unequivocally as a nova; CG CMa,
with a maximum recorded by van Hoof (1948); and WX Cet, announced
as a nova by Strohmeier (1963). A search of Harvard photographs
by Gaposchkin (unpublished) has revealed several earlier maxima of
WX Cet, which I believe to be a U Gem star; CG CMa is denoted N?
in the Moscow catalogue. Further study might find KT Mon to be
recurrent (and hence fainter than a classical nova), and CG CMa
might prove to be a U Gem star or a recurrent nova. A number of
stars, once thought to be novae and so faint at maximum as to
place them at large distances from the galactic center, have been
weeded out from the earlier lists (Table 4).

Table 4

Stars once Announced to be Novae

Present Class	Stars
U Geminorum	KY Ara, WX Cet, DM Lyr, UW Per?, SW Vul
Z Andromedae	CM Aql, FN Sgr, V 1017 Sgr?
Mira	V 607 Aql, V 407 Cyg, UV Eri, SU Lyr, HN Lyr, V 939 Sgr
Missing BD stars	W Ari, SU Ari, SY Gem, VZ Gem, U Leo, V 728 Sco
Not variable?	X Vir
False image	Nova Psc 1907
Misidentification?	V 529 Ori?
Others	Eta Car, P Cyg, FN Ori, V 605 Aql

The frequency of r sin b is illustrated in Figure 2; clearly
the novae are concentrated toward the galactic plane, and conform
to an intermediate rather than a spherical population. Five accepted novae appear to be more than 2000 parsecs from the plane:
BD Pav (-2770), RW UMi (+2720), CT Ser (+3670), X Ser (+5050) and
the recurrent VY Aqr (-8190). If WX Cet were a classical nova it

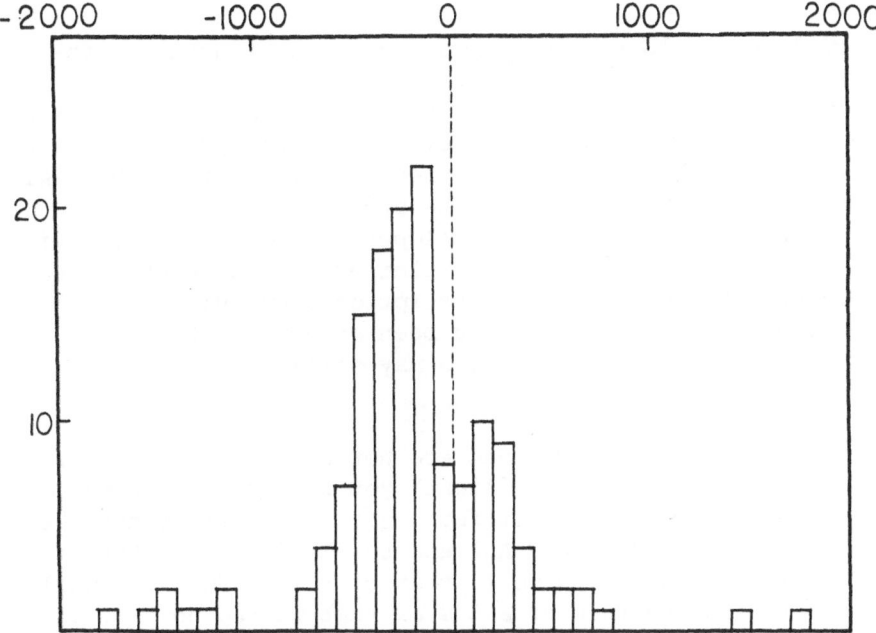

Figure 2. Frequency of distance from the galactic plane for novae.
Ordinates are numbers within 100 parsec intervals; abscissae are
parsecs.

would be even farther from the plane (-39,300); as a U Gem star
it would be a little more than 1000 parsecs below the plane.
Possibly some of the other stars mentioned are recurrent novae or
U Gem stars and would repay a search.

The distribution of novae in low latitudes shows clear evidence of the effects of obscuration. Most of the novae within
+ 100 parsecs from the galactic plane must be concealed by obscuration, especially in positive latitudes. The median observed
r sin b is 220 parsecs, but the true value must be much less on
account of the selective effects of absorption in low latitudes.

THE SPECTRUM DURING THE OUTBURST

The characteristic spectrum during outburst identifies a nova
as such. At maximum it is commonly described as being similar to
that of a luminous A or F star; after an interval permitted bright
lines develop, accompanied by violet-shifted absorption components,
often multiple, with complex changes of radial velocity. Still
later, the "nebular" spectrum of forbidden lines develops, and
finally dies away, leaving the post-nova spectrum showing a blue
continuum, bright lines, and sometimes the spectrum of a second

red component.

I will not yield to the temptation of expatiating on the
beautiful subject of the development of the nova spectrum. I
confine myself to a story that may be new to some of you. It was
told to me by Miss Cannon. When Nova GK Persei appeared in 1901,
she hurried with the coming of darkness to superintend the taking
of photographs of the spectrum. She stood in the darkroom, waiting
for the plate to come from the developer. She held the dripping
spectrum to the light. Then she exclaimed: "You've photographed
the wrong star!" For it was an absorption spectrum. Up to that
time, only bright-line spectra had been observed for novae.
History could repeat itself: novae may still have surprises in
store for us.

The maximal spectra of novae are not all alike. That of
V 1500 Cyg seems to be the "earliest" yet recorded, with the one
of GK Per that made history not far behind. Spectra range through
the early A spectrum of V 603 Aql and the middle F spectra of RR
Pic. The "latest" known to me is that of V 1148 Sgr, the well-
marked absorption spectrum of a K star that wasted no time in
developing the characteristic bright lines, so it must have been
caught near maximum brightness. This nova is one of the three
that have been associated with globular clusters, the other two
being T Sco and the unnamed nova in Messier 14, announced by
Sawyer Hogg and Wehlan (1964,1965). The spectra of the two latter
were not recorded; possibly novae in globular clusters are exceptional.

One more star calls for brief comment: V 605 Aql which was
of about eleventh magnitude (pg) in 1919. Herbig (1958b) gives a
magnitude of about 20 (pg) from the Palomar Atlas. Figure 3 shows

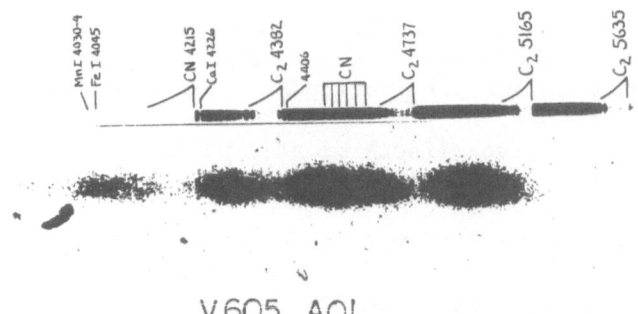

Figure 3. Spectrum of V 605 Aquilae, compared with that of the
standard R 5 star in the Morgan-Keenan Atlas. See text.

the spectrum, taken by Lundmark on a slitless plate at the Crossley
telescope of the Lick Observatory on September 22, 1921. Lundmark

described it as of class RO, a classification with which I have
no quarrel. It matches closely with the standard R5 star
HD 52432 in the Morgan-Keenan Atlas; in addition to the Carbon
bands the Ca I line at 4227 is clearly visible. Certainly it is
not a bright-line spectrum, as would be expected at this stage of
the light curve, more than two years after brightening, if the
star were a conventional nova. It does not seem to resemble any
star that I know, considering its erratic variation of large range
and its possible association with a planetary nebula, pointed out
by van den Bergh (1971). Bidelman (1973) suggests a relation to
the RCrB class on the basis of another spectrum taken by Lundmark
in September, 1921. A similar suggestion had in fact been made
by Ludendorff (1922).

    The absorption spectra of novae at maximum light, as mentioned
earlier, resemble those of luminous stars of spectral classes B,
A, F, and (in one case) K. They are not, however, exactly like
their static counterparts, and (considering the violent disturbance
that underlies them) it is surprising that they should be as like
as they are. The application of conventional methods to derive
abundances from such spectra is fraught with difficulties and
dangers, and the results should be accepted with due reserve.
Studies of several slow novae, notably DQ Her and HR Del, have
led to the conclusion that C, N and O are more abundant than in
the sun; see the summary by Antipova (1974), the discussion by
Mustel and Antipova (1971), and also the evaluations of composition from the bright-line spectra by Pottasch (1959,1967).

MINIMAL BRIGHTNESS AND RISE

    The view has been widely accepted that the pre- and post-nova
stages are essentially similar, and that the brightness before
and after outburst is the same. The rise in brightness has been
regarded as abrupt and extremely rapid. Robinson (1975) has
argued that this behavior is not characteristic of all novae.

    Before its outburst, V 446 Her varied with a range of nearly
4 magnitudes. Since recovery it has seemed to vary only slightly,
with a mean magnitude similar to the mean magnitude before eruption. It is as though the star underwent a series of disturbances
with a rise time of the order of ten days, as a prelude to the
main outburst. Something similar seems to have been going on in
the pre-eruption light of V 3890 Sgr (see Robinson's Figure 10,
where the star is designated Nova Ser 1962), but the post-nova in
this case has not been investigated. Two other novae, both very
different from the above, showed analogous behavior. The light
curve of the blue component of T CrB underwent a preliminary semi-
periodic disturbance with a period near to that of the binary
system, before the outburst of 1946. The amplitude increased with

time, and culminated in the main outburst. Also the very slow
nova RR Tel, which had displayed small semi-periodic variations at
minimum for more than 10,000 days, varied with a period of 387
days for more than 2000 days before the main outburst, and this
period persisted during the eruption itself. As RR Tel has not
yet completed its decline, a comparison of pre- and post-nova is
not yet possible. The extremely well-observed photographic light
curve by Mayall (1949) shows that minimal light had fallen to the
thirteenth magnitude at JD 2413000, reached fourteenth magnitude
near JD 2416000, began to fluctuate by MD 2426000, and fell to
near the sixteenth magnitude near JD 2428000; after this the well-
marked periodic fluctuations began to mount up, culminating in
the major outburst more than 2000 days later. There is thus evidence that at least four novae, differing greatly in character,
were undergoing premonitory disturbances long before the main outburst. The two last named are binaries, giant M stars with blue
companions. In neither case did the maximal spectrum show evidence
of the M star: that of T CrB was of very early type, that of RR
Tel resembled an F supergiant.

The light curves of V 446 Her, V 3890 Sgr and T CrB were
those of fast novae. Another fast nova that should be mentioned
in connection with pre- and post-outburst behavior is GK Per. The
pre-eruption variations as illustrated by Robinson showed fluctuations several hundred days before the main outburst. Figure 4

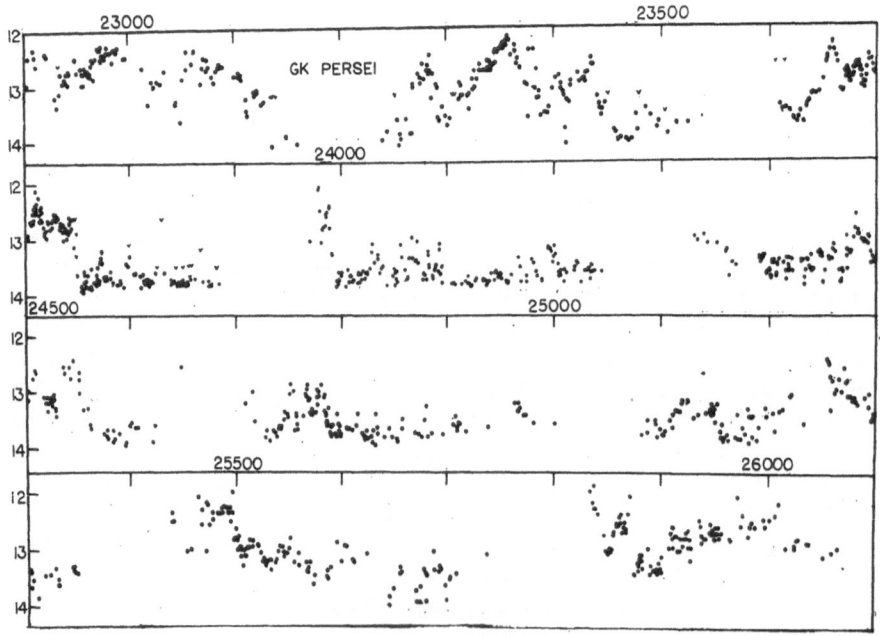

Figure 4. Post-maximum variations of GK Persei as observed visually by Steavenson and Peek (see text). Ordinates are visual
magnitudes; abscissae are Julian Days.

shows some of the beautiful and detailed visual observations of
the post-nova variations made by Steavenson (1921,1923,1924,1925,
1926,1927,1928,1929,1930,1931) and Peek (1924,1925,1926,1927,1928,
1931). These curves emphasize the continuous fluctuations of
large range (up to 1.5 magnitudes) with cycles from tens to
hundreds of days, on which (as these observers noted) were superimposed rapid changes of brightness in less than a day. Besides
throwing light on the continuous activity of the post-nova, the
curves should remind us that caution should be used in assigning
a minimal magnitude to a nova on the basis of a few necessarily
scattered observations, as is done for many pre-novae.

The second point made by Robinson is that the rise is not
abrupt for all novae. This point has already been made for T CrB,
RR Tel, and RR Pic. He points out that H 533 Her and LV Vul had
had small but significant increases in luminosity in the few
years before the outburst, and suggests that this may also have
been the case with CP Lac, BT Mon and GK Per. Recently V 1500 Cyg
has been found to show a similar phenomenon. All six of these
stars were fast novae. When we recall that very few novae have
been intensively observed before outburst, we may consider that
a pre-outburst disturbance may not in fact be at all exceptional.

DWARF NOVAE AND NOVA-LIKE VARIABLES

A census of the recognized dwarf novae comprises 284 stars:
255 of U Geminorum, 29 of Z Camelopardalis type. Ranges are
recorded for nearly half of these (Table 5), 125 U Gem and 22 Z
Cam stars. The largest range recorded (9 magnitudes) is that of
Al Com, with an observed cycle of 1230 days and a spectrum described as resembling that of U Gem; WX Cet, with a range of 9
magnitudes is thus not outside the limits of the class.

The dwarf novae are formally defined as having ranges up to
6 magnitudes. Their spectra differ from those of classical novae:
at maximum they appear to have a strong blue continuum, and often
very wide absorption lines of H and He I. At a minimum, when observed, they show various combinations of the bright-line spectrum
of a hot star and the absorption spectrum of a main sequence star
of class G or K. Thus, when compared to classical novae, their
ranges are smaller, and their maximal spectra ordinarily lack
bright lines. However, there is some overlap: SW UMa shows weak
emission even at maximum. We may consider that novae eject
material at the outburst, and that dwarf novae do not; but the
distinction is not clear-cut.

Dwarf novae are generally regarded as conforming to the relation between amplitude and length of cycle. But this is true
only in the most general way. Observed amplitude is plotted

Table 5

Range	U Gem stars	Z Cam stars
	Dwarf Novae: Frequency of Range	
1.0 to 1.49	7	1
1.5 to 1.99	6	3
2.0 to 2.49	12	6
2.5 to 2.99	16	2
3.0 to 3.49	11	6
3.5 to 3.99	18	1
4.0 to 4.49	24	2
4.5 to 4.99	13	1
5.0 to 5.49	8	0
5.5 to 5.99	7	0
6.0 to 6.49	1	0
6.5 to 6.99	1	0
9.0 to 9.49	1	0

against length of cycle in Figure 5, for cycles up to 160 days.

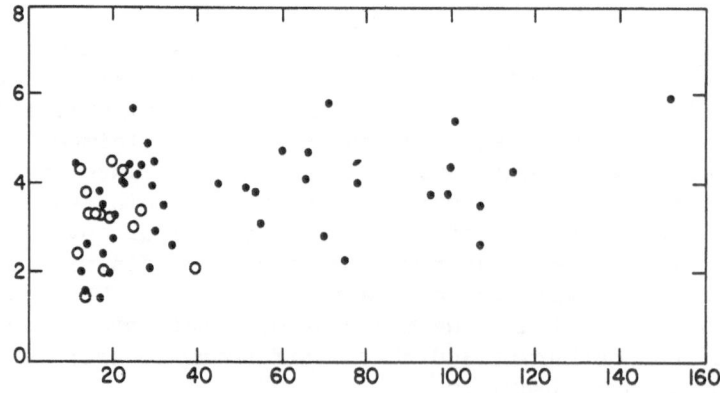

Figure 5. Range (magnitudes) and cycle (days) for U Geminorum stars (dots) and Z Camelopardalis stars (circles); cycles over 160 days omitted.

Only the most optimistic eye would discern a correlation. That
the Z Cam stars, which have on the average the smallest ranges,
and also the smallest cycles is a matter of class definition. The
ten U Gem stars with cycles greater than 160 days have ranges
from 2.2 to 9 magnitudes: there is a larger proportion of large
ranges for long cycles, but certainly no close correlation.

To complete the census of novae we must include the stars
catalogued as "nova-like," a class that has in the past included
stars like Z And that have cyclic nova-like outbursts, stars that
show small, erratic brightenings (such as GK Persei at minimum),
and stars whose spectra have nova-like characteristics. Some
that have been included with novae in the past are now relegated
to the new S Doradus or γ Cassiopeiae classes. Most of the rest
belong in the Z And class, binary stars with a cool giant or
supergiant component and a blue star that is presumably the seat
of the outburst. Many of them must eject material at some of
their light maxima, and are to be considered novae of a sort.
The recurrent nova T CrB, a spectroscopic binary with period
227.6 days and the slow nova RR Tel link these objects with the
classical novae. The peculiar Mira star R Aqr links them with
the red variables of long period. Table 6 contains a list of Z
And stars. The observed periods listed in the table may be regarded as orbital, except those of the intrinsic variables R Aqr
and o Ceti.

NOVAE AS BINARY SYSTEMS

That classical and dwarf novae are binary systems has been
amply demonstrated within the past twenty years. Save for T CrB
the recognized periodic binaries consist of a main sequence star
and star of low luminosity and high temperature which is the seat
of the outburst. Along with this discovery has come the recogni-
tion of a number of systems that resemble novae at minimum both
in spectrum and in photometric properties. These cataclysmic
variables, considered as a group, have complex photometric behavior: (1) the main outburst or outbursts; (2) erratic fluctuations,
some amounting to several magnitudes (V 446 Her, GK Per) with a
time scale of tens or hundreds of days; (3) rapid, irregular
flickering with a time scale of hours or minutes; (4) rapid co-
herent oscillations with a time scale of tens of seconds; (5) ex-
trinsic variations with the orbital period, compatible with
eclipses.

A model of the cataclysmic variable is described by Robinson
(1976), who considers that it involves a white dwarf and a late
type star that fills its Roche lobe and is transferring mass to
the blue component. The latter is surrounded by an accretion
disk, and the point at which the accreted mass impinges on the

Table 6

Z Andromedae Type and Related Stars

Name		M	m	Spectrum	Period Orbital	Intrinsic	Remarks	Reference
Z	And	8.0	12.4	M2III + Beq	694			
R	Aqr	5.8	11.5	M5e-8.5e + B2eq		387	Mira star	
CM	Aql	13.0	16.4	like Z And				1
TX	CVn	9.3	11.6	K2 + B5eqv				2
RT	Car	11.0	11.4	M2Ia + (OB)				
o	Cet	2.0	10.1	gM5e – gM9e		331.65	Mira star	
VZ	Cet	9.5	12	Beq	vis.bin, 100Y		var. on 15 min scale; rare short 2 min flares	3
BF	Cyg	9.3	13.4	Bep + gM4	750			4
CI	Cyg	10.7	13.1	Bep + gM4	856		eclipsing?	
YY	Her	11.7	[13.2]	M2ep + 0				
V443	Her	12.39	12.63	M3ep + 0				
RW	Hya	10.0	11.2	gMep	876			5
SS	Lep	4.82	5.06	A0Veq + gM1	276:		Infra-red excess	6
SY	Mus	11.3	12.3	Mep, composite	621			7

Table 6 (cont.)

Name	M	m	Spectrum	Period Orbital	Period Intrinsic	Remarks	Reference
AR Pav	10.7	12.7	Beq + cF + M	605		Eclipsing	8
AG Peg	6.0	9.4	WN6 + M1-3 II-III	830			9
AX Per	10.8	13.0	gM3ep + hot star	600-800			
V1017 Sgr	6.2	14.4	G5IIIep + hot star				
V2416 Sgr	14.6	[17.6				10 year P? planetary nebula?	10
V2601 Sgr	14.0	15.3	M6e, composite			Bright H, HeI, HeII	
V2905 Sgr	10.0	14.6	Beq? + M?				

References, Table 6
----

1. Mumford, 1956.
2. Münch and Morgan, 1953; Humphreys, Strecker and Ney, 1972.
3. Warner, 1972b.
4. Aller, 1954.
5. Mayall, 1957.
6. Wright, 1957; Widing, 1966; Allen and Ney, 1972.
7. Henize, 1952.
8. Thackerary, 1959; Andrews, 1974; Thackeray and Hutchings, 1974.
9. Boyarchuk, 1966; Cowley, 1973.
10. Herbig, 1969.

accretion disk forms a bright spot.

The complex photometric variations have been identified with
the interplay of these components. (1) The main outburst for
novae has been ascribed to a thermonuclear runaway in the envelope
of the white dwarf; the eruption in dwarf novae can be attributed
to a sudden expansion from the disk around the white dwarf,
perhaps with a contribution from the white dwarf itself. (3) and
perhaps (2) have been ascribed to rapid changes in the bright
spot, which is a shock front produced by the impact of accreted
material on the disk. (4) are seen during and immediately after
eruptions, and have been attributed to non-radial oscillations
of a white dwarf or the disk associated with it. (5) are caused
by eclipses of the hot star, disk, and spot, or some combination
of them.

Not all of this photometric behavior is shown by all cataclysmic variables. (1) is displayed by novae and dwarf novae;
(2) and (3) are probably typical of all, (4) is probably associated with all outbursts, and (5) depends on the orientation of
the system: EY Cyg while undoubtedly binary does not eclipse.
The above identification of the sources of the various types of
photometric behavior is schematic and over-simplified.

The published information on the classical novae, the dwarf
novae, the "once or future novae" and (for completeness) the ZZ
Ceti stars, "variable white dwarfs," is summarized in Tables 7 to
10.

The four tables have much in common. Orbital periods for the
classical novae, dwarf novae, the ex-novae and the ZZ Ceti star
all have similar ranges, and show no relation to the morphology
of the outburst. The two fast novae (V 603 Aql and GK Per) have
respectively one of the shortest and one of the longest orbital
periods. If the periodic variations of V 1500 Cyg, recorded by
Tempesti (1975,1976), Ambruster et al. (1976), Chia et al. (1976),
and Semeniuk (1975), Rosino and Tempesti (1976), and Semeniuk
et al. (1976) are interpreted as evidence of orbital motion in a
binary system, they suggest similarity with V 603 Aql. The orbital
period for dwarf novae bears no relation to the length of the outburst cycle.

Only the hot star is recorded for the binary systems of the
shortest period, as has long been known. This gives an idea of
the probable periods of stars with recorded minimal spectra, not
yet known to be binaries. At minimum the following stars show
evidence of the main sequence component in their spectra: UU Aql,
EY Cyg, AH Her, and EZ Peg; they may be expected to have periods
greater than 6 hours.

Table 7

Classical and Recurrent Novae

Name	Type	M	m	Ecl.	S.B.	Period Orbital d	Puls. s	Sp.	Ref.	Outburst
V 603 Aql	Ma	-1.1	12.8		*		..	Be	1	1918
T Aur	Nb	4.1	15.8	*		0.2043786	..	..	2	1891
T Crb	Nr	2.0	10.8		*	227.6	..	Be+gM3+Q	3	1866,1946
HR Del	Nb	3.5?	12?		*	0.1913515?	..	Q	4	1967
DQ Her	Nb	1.3	15	*	*	0.19362070	71	Be + Q	5	1934
DI Lac	Na	4.3	14.9				..	Be + Q	6	1910
RS Oph	Nr	5.2	12.3				..	Ocp+M2ep	7	1898,1933,1958, 1967
V 841 Oph	N	4.2	13.1				..	Be	8	1848
RR Pic	Nb	1.2	12.8	*		0.1450255	..	..	9	1925
GK Per	Na	0.2	14.0		*	0.684722	..	Be+K2IVp+Q	10	1901
WY Sge	N	5.4	19.5				..	..	11	1782
WZ Sge	Nr	7.0	15.5	*	*	0.056688	..	Be	12	1913,1946
RR Tel	Nc	6.5	16.5				..	M5III+Q	13	1944

References, Table 7

1. Kraft, 1964.

References, Table 7 (cont.)

2. Walker, 1963b; Mumford, 1967a.

3. Kraft, 1958; Weber, 1960.

4. Szumiejko, 1976.

5. Walker, 1961; Kraft, 1964; Nather and Warner, 1969; Hubbard et al., 1972; Herbst, Hesser and Ostriker, 1974; Kiplinger and Nather, 1975.

6. Kraft, 1964.

7. Wallerstein, 1963; Barbon et al., 1968; Tempesti, 1975a.

8. Kraft, 1964.

9. van Houten, 1966; Mumford, 1971; Vogt, 1975.

10. Kraft, 1964; Paczynski, 1965b; Mumford, 1971b; Gallagher and Oinas, 1974.

11. Warner, 1971: 10-minute fluctuation?

12. Kraft et al., 1962; Krzeminski, 1962; Krzeminski and Smak, 1971; Mumford, 1971; Warner and Nather, 1972b.

13. Feast and Glass, 1974.

Table 8

Dwarf Novae

Name	Type	Max.	Min.	Ecl.	S.B.	Period Orbital d	Period Puls. s	Cycle d	Spectrum	Ref.
RX And	Z Cam	10.3	13.6		*	0.21173		14.1	Be	1
AR And	UG	12.8	17.3					30	like U Gem	
AE Aqr	Z Cam?	10.4	12.0		*	0.4116550			Be? + dKO	2
UU Aql	MG	11.0	16.8					71.3	Gep	
SS Aur	UG	10.5	14.8		*	0.180556		558	Be	3
Z Cam	Z Cam	10.2	14.5	*	*	0.2878	16+	22	Be + G?	4
SY Cnc	Z Cam	10.6	14.0				24.62	27.3	like Z Cam	5
BV Cen	UG	10.5	14.2	*		0.1580		99.3		6
V436 Cen	UG	11.9	15.4	*		0.064028	19 to 20	..		7
WW Cet	UG	9.3	[16.3		*	0.159722			like SS Cyg	8
Z Cha	UG	11.5	16.3	*		0.074502	27.67	66		9
SS Cyg	UG	8.2	12.1		*	0.276		51.6	Be + dG5	10
EY Cyg	UG	11.4	15.7					1490:	Be + KOV	11
U Gem	UG	8.8	14.2	*	*	0.176906		101.8	Be	12
AH Her	Z Cam	10.2	14.7				31.55	19.8	like U Gem,G: at minimum	13

Table 8 (cont.)

Name		Type	Max.	Min.	Ecl.	S.B.	Period Orbital d	Period Puls. s	Cycle	Spectrum	Ref.
EX	Hya	UG	11.4	14.1	*	*	0.06823		465:	like U Gem, double emission	14
VW	Hyi	UG	8.5	13.4	(*)		0.0742711	28–34	28.7	no eclipse, hump	15
WX	Hyi	UG	11.5	14.73						continuous, H in emission	
T	Leo	UG	10	15.4						wide bright H, HeI, CaII	
X	Leo	UG	11.5	15.5					22.7	continuous at min, no emission	
AY	Lyr	UG	12.6	17.0					24	G:	
CY	Lyr	UG	13.2	17.0					17	Pec.	
IK	Nor	UG?	12.9	15.5					34	Pec. (O)	
CN	Ori	Z Cam	11.6	14.8				24.5	19.2	Pec.	16
RU	Peg	UG	9.0	13.1		*	0.370833		65.7	Be + G8 IVn	17
EZ	Peg	UG	9.5	10.5						G5Ve(+B)	
KT	Per	Z Cam	10.7	15.0				26.73	12		18
TY	Psc	UG	12.5	16						Pec.:	
UZ	Ser	UG	12.0	16.7					26.1	Pec.	
SU	UMa	UG	11.0	14.49					17.4	Max.,A5–7;min.,faint continuum,wide H,HeI?	

SW	UMa	UG	10.6	16	459	Max.,strong H;min, faint continuum, bright H
CH	UMa	UG	10.7	15.9	400:	Max.,continuum,no features
TW	Vir	UG	11.8	16	26	Pec.

References, Table 8

1. Kraft, 1962; Robinson, 1973; Szkody, 1974a.
2. Crawford and Kraft, 1956; Walker, 1965; Payne-Gaposchkin, 1969.
3. Herbig, 1960; Kraft, 1962.
4. Kraft, 1962; Mumford, 1963; Kraft, Krzeminski and Mumford, 1967; Robinson, 1973c.
5. Herbig, 1950; Robinson, 1973b.
6. Mumford, 1971.
7. Warner, 1975b.
8. Herbig, 1962; Kraft, 1964.
9. Mumford, 1971; Warner, 1974.
10. Kraft, 1962; Walker and Chincarini, 1968; Smak, 1969; Walker and Reagan, 1971; Holm and Gallagher, 1974; Szkody, 1974b.
11. Kraft, 1962.
12. Kraft, 1962; Pacynski, 1965a; Krzeminski, 1965; Mumford, 1969; Mumford, 1970; Warner and Nather, 1971.
13. Elvey and Babcock, 1943; Robinson, 1973a.
14. Mumford, 1964; Mumford, 1965; Mumford, 1967b; Mumford, 1971; Warner, 1972a; Warner, 1973.
15. Warner and Harwood, 1973; Warner and Brickhill, 1974; Vogt, 1974; Warner, 1975.
16. Warner and Robinson, 1972.
17. Kraft, 1962; Luyten and Hughes, 1965.
18. Robinson, 1973.

Table 9

Ex-novae or Potential Novae

Name	Max.	Min.	Ecl.	S.B.	Period Orbital d	Puls. s	Spectrum and Remarks	Ref.
TT Ari	10.2	11.8	*	*	0.2658		B	1
AM CVn	13.94	13.98	*		0.01216	115	DBp	2
QU Car	11.3	11.6					Ape; resembles V818 Sco?	3
EM Cyg	11.9	14.4	*	*	0.290909	18–25	like V Sge	4
V751 Cyg	13.8	16.3					like UX UMa; continuous eruption	5
MV Lyr	10.5	14.0			0.08::		Op; flickering	6
V426 Oph	11	13.8					Pec., diffuse emission, H,HeI	7
VV Pup	14.6	17.1	*	*	0.06975		Pec., strong emission, H,HeII	8
V Sge	9.5	13.9	*	*	0.514195		WN5 + dG	9
V818 Sco	11.1	14.1	*?	*	0.787313		Bright H,HeII,CIII,NIII; X-ray source; variable radio source	10
VY Scl	12.9	18.5					OB ce; flickering, continuous eruption	11
VZ Scl	15.6	18.1	*	*	0.14462220		Be; continuous eruption	12
V471 Tau	9.40	9.71	*		0.52118346	none	KOV + D; no rapid variations	13
RW Tri	13.5	16.04	*	*	0.23188		like UX UMa; reverse P Cyg effect	14
UX UMa	12.7	13.8	*	*	0.19667128	28.5–30	0	15

Table 9 (cont.)

Name	Max.	Min.	Ecl.	S.B.	Period Orbital d	Puls. s	Spectrum and Remarks	Ref.
AN UMa	13.8	14.4	*		0.0796894		like V Sge	16
CoD -42°14462				*		29.0	continuous eruption; binary	17

References, Table 9

1. Smak and Stepien, 1969; Cowley, 1975.
2. Greenstein and Matthews, 1957; Smak, 1962; Burbidge, Burbidge and Hoyle, 1967; Faulkner, Flannery and Warner, 1972; Warner and Robinson, 1972; Krzeminski, 1972.
3. Stephenson et al., 1968; Schild, 1969.
4. Burbidge and Burbidge, 1955; Mumford and Krzeminski, 1969; Faulkner, 1971; Warner and Nather, 1971, 1972; Robinson, 1974.
5. Herbig, 1958a; Robinson, Nather and Kiplinger, 1974.
6. Walker, 1954; Greenstein, 1954; Weber, 1960; Walker, 1966.
7. Herbig, 1960a.
8. Herbig, 1959, 1960b; Warner and Nather, 1972c.
9. Herbig, Preston, Smak and Pacyniski, 1965.
10. Robinson and Warner, 1972; Cowley and Crampton, 1975; Gottlieb et al., 1975.
11. Burrell and Mould, 1973; Warner and van Citters, 1974; Nather and Kiplinger, 1974.
12. Walker, 1956; Chavira, 1958; Greenstein, 1966; Krzeminski, 1966; Warner and Thackeray, 1975.
13. Warner, Robinson and Nather, 1971.
14. Walker, 1963.
15. Walker and Herbig, 1954; Nather and Warner, 1972; Robinson, Nather and Kiplinger, 1974; Nather and Robinson, 1974.
16. Shugarov, 1975; Mumford, 1976.
17. Wegner, 1972; Warner, 1973a; Robinson, Nather and Kiplinger, 1974; Hesser et al., 1974; Cowley et al., 1976.

Table 10

ZZ Ceti Stars

Name		Max.	Min.	Spectrum	Remarks	Ref.
ZZ	Cet	14.09	14.11	DA	$P_1=212^S$, $P_2=273^S$	1
GP	Com	15.69	15.9	DB	spectroscopic binary? $P=0^d.261$; flicker	2
CY	Leo	15.8:	16.84	DC	$600^S,822^S,1638^S$; flares	3
ZZ	Psc	13.1	13.2	DA	$816^S,612^S$ alternate	4
V3885	Sgr	10.30	10.44		Wide H,HeI lines	5
V411	Tau	14.09	15.28	DA	$P_o=0^d.008636, P_1=0^d.005721$; $P=0^d.135$	6

References, Table 10

1. Luyten, 1949; Lasker and Hesser, 1971.

2. Warner, 1972c; Richer et al., 1973.

3. Lasker and Hesser, 1969; Warner et al., 1970; Greenstein, 1970.

4. Warner, 1973a; Hesser, Lasker and Osmer, 1974.

5. Wegner, 1972.

6. Warner and Nather, 1972a; Fitch, 1973.

SUMMARY

     I have attempted a critical survey of the information neces-
sary for a decision as to the kinds of objects that become, or
can become, novae. This has involved definitions and lists of
classical novae, dwarf novae, ex-novae and nova-like (Z Andromedae) stars. The first group undergoes outbursts with ejection
of material; the others (with exceptions such as Z And, R Aqr
and perhaps IK Nor) do not. The invariable requirement is the
association of a hot star of low luminosity (classical and dwarf
novae) or a hot subdwarf (Z Andromedae stars) with a star of late
type. I have included the ZZ Ceti stars in a final table, because
V 411 Tau, though regarded as a variable white dwarf, displays
variations of type (3), and GP Com is a possible binary. It is
my belief that all these stars are symbiotic (an extension of the
application of this term, originally applied to systems with one
"giant" component, to systems where mutual influence affects behavior). I have avoided theoretical speculations, and have aimed
at referring to facts that seem relevant to a decision as to the
physical nature of cataclysmic variables.

     It has been suggested many years ago that the W Ursae Majoris
stars may be the precursors of the binaries that can become novae.
In conclusion I illustrate (Figure 6) the frequency of period of
the 420 known W Ursae Majoris stars and the stars of Tables 7 to 10
whose periods are known. It seems that the W Ursae Majoris stars

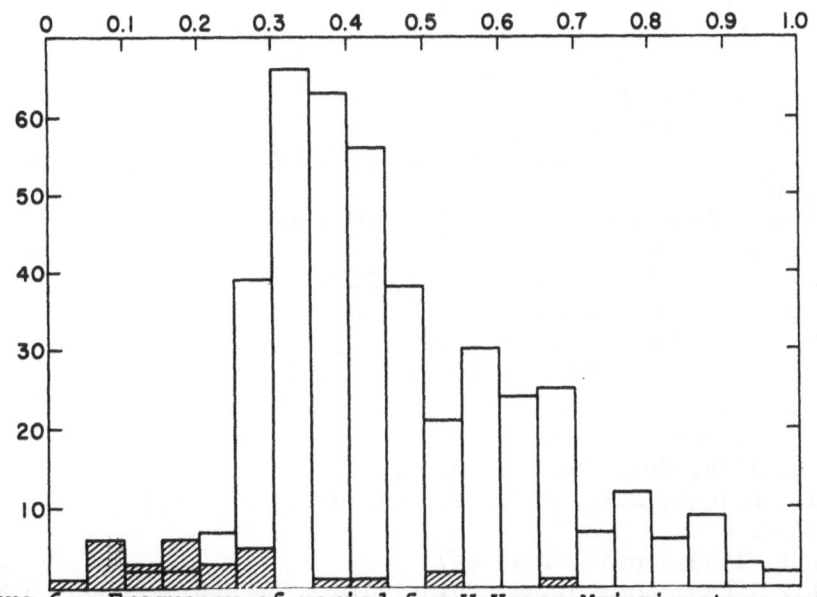

Figure 6. Frequency of period for W Ursae Majoris stars and cataclysmic variables of known period (shaded area). Ordinates are
numbers of stars, abscissae are days.

with periods less than $0^d.2$ present an interesting field of en-
quiry.   It should be noted that W UMa itself has been observed
by Kuhi (1964) to have undergone a flare of amplitude $1^m.5$ and
7 minutes' duration; this contact binary, period $0^d.33364667$,
shows two spectra of class dF8p. Even the Algol variable U Cep
has displayed an outburst, recorded by Plavec and Polidan (1975)
and Batten et al. (1975), and ascribed by the latter to a rotating
disk.

Finally we may consider the undiscovered ex-novae. On the
crude but widely-accepted assumption that fifty novae brighter
than the seventh apparent magnitude at maximum occur in our
Galaxy every year, and that the cycle of the classical nova is of
the order of 10^4 years, there should be more than 10^5 ex-novae
brighter than the nineteenth apparent magnitude. If classical
novae, as Robinson (1976) suggests, have but one outburst, they
would be far more numerous. To judge by the observed distribution of the known novae, these ex-novae should lie preferentially
in the direction of the galactic center. They would be rapid
erratic variables of small range, many should be blue stars, and
in addition a substantial fraction should be eclipsing variables
of very short period. A systematic attempt to find such stars in
the 400 square degrees around the position of the galactic center
would be of great interest.

BIBLIOGRAPHY

Allen, D.A., and Ney, E. 1972. Obs., 92, 47.
Ambruster, C.W., Blitzstein, W., Hull, A.B., and Koch, R.H. 1976,
    Bull. A.A.S., 8, 329.
Andrews, P.J., 1974. M.N.R.A.S., 167, 635.
Antipova, L.I., 1974. Highlights, 3, 501.
Barbon, R., Mammano, A., and Rosino, L., 1968. I.A.U. Colloquium
    No. 4, 257.
Batten, A.H., Fisher, W.A., Baldwin, B.W., and Scarfe, C.D. 1975.
    Nature, 253, 174.
van den Bergh, S. 1971. P.A.S.P., 83, 819.
Bidelman, W.P. 1973. Bull. A.A.S., 5, 442.
Boyarchuk, A.A. 1966. A Zh., 43, 976.
Burbidge, E.M. and Burbidge, G.R. 1955. Ap.J., 118, 349.
Burbidge, G., Burbidge, M., and Hoyle, F. 1967. Ap.J., 147, 1219.
Burrell, J.F., and Mould, J.R. 1973. P.A.S.P., 85, 627.
Chavira, E. 1958. Bol. Ton. y Tac., 2, 17.
Chia, T.T., Milone, E.F., Robb, R., and Wytsma, R. 1976. Bul. A.A.S.,
    8, 329.
Cowley, A.P. 1973. Ap.J., 184, 687.
Cowley, A.P., Crampton, D., Hutchins, J.B., and Marlborough, J.M.
    1975, Ap.J., 195, 413.
Cowley, A.P., and Crampton, D. 1975. Ap.J., 201, L 65.

Cowley, A.P., Crampton, D., and Hesser, J.E. 1976. This meeting.
Crawford, J.A., and Kraft, R.P. 1956. Ap.J., 123, 44.
Elvey, C.T., and Babcock, H.W. 1943. Ap.J., 97, 412.
Faulkner, J. 1971. Ap.J., 170, 99 L.
Faulkner, J., Flannery, B., and Warner, B. 1972. Ap.J., 175, 79 L.
Feast, M.W., and Glass, I.S. 1974. M.N.R.A.S., 167, 81.
Fitch, W.S. 1973. B.A.A.S., 5, 17.
Gallagher, J.S., and Oinas, V. 1974. P.A.S.P., 86, 952.
Gaposchkin, S. 1954. A.J., 59, 199.
Gottlieb, E.W., Wright, E.L., and Liller, W. 1975. Ap.J., 195,
    L 33.
Greenstein, J.L. 1954. P.A.S.P., 66, 79.
Greenstein, J.L. 1966. Ap.J., 144, 496.
Greenstein, J.L. 1970. Ap.J., 162, L 55.
Greenstein, J.L., and Matthews, M.S. 1957. Ap.J., 126, 14.
Henize, K.C. 1952. Ap.J., 115, 133.
Herbig, G.H. 1950. P.A.S.P., 62, 211.
Herbig, G.H. 1958a. Ap.J., 128, 259.
Herbig, G.H. 1958b. P.A.S.P., 70, 605.
Herbig, G.H. 1959. A.J., 64, 128.
Herbig, G.H. 1960a. Ap.J., 131, 632.
Herbig, G.H. 1960b. Ap.J., 132, 76.
Herbig, G.H. 1962. H.A.C., 1576.
Herbig, G.H. 1969. Lick Contr., 299.
Herbig, G.H., Preston, G., Smak, J., and Pacynski, B. 1965. Ap.J.,
    141, 617.
Herbst, W., Hesser, J.E., and Ostriker, J.P. 1974, 193, 679.
Hesser, J.E., Lasker, B.M., and Osmer, P.S. 1974. Ap.J., 189, 315.
Holm, A.V., and Gallagher, J.S. 1974. Ap.J., 192, 425.
van Hoof, A. 1948. Louvain Cobtr., 109, 14.
van Houten, C.J. 1966. B.A.N., 18, 439.
Hubbard, W.B., Warner, B., Peters, W.L., and Nather, R.E. 1972.
    M.N.R.A.S., 159, 321.
Humphreys, R.M., Strecher, D.W., and Ney, E.P. 1972. Ap.J., 172,
    75.
Kiplinger, A.L., and Mather, R.E. 1975. Nature, 255, 125.
Kraft, R.P. 1958. Ap.J., 127, 625.
Kraft, R.P. 1962. Ap.J., 135, 408.
Kraft, R.P. 1964a. Ap.J., 139, 457.
Kraft, R.P. 1964b. Conference on Faint Blue Stars, 82.
Kraft, R.P., Matthews, J., and Greenstein, J.L. 1962. Ap.J., 136,
    312.
Kraft, R.P., Krzeminski, W., and Mumford, G.S. 1969. Ap.J., 158,
    589.
Krzeminski, W. 1962. P.A.S.P., 74, 66.
Krzeminski, W. 1965. Ap.J., 142, 1051.
Krzeminski, W. 1966. I.B.V.S., 160.
Krzeminski, W. 1972. A.A., 22, 387.
Krzeminski, W., and Smak, J. 1971. A.A., 21, 133.
Kuhi, L. 1964. P.A.S.P., 76, 430.

Kukarkin, B.V., Kholopov, P.N., Efremov, Yu.N., Kukarkina, N.P., Kurotchkin, N.E., Medvedeva, G.I., Perova, N.B., Fedorovich, V.P. and Frolov, M.S., 1969. General Catalogue of Variable Stars, Sternberg State Institute, Moscow.

Kukarkin, B.V., Kholopov, P.N., Efremov, Yu.N., Kukarkina, N.P., Kurochkin, N.E., Medvedeva, G.I., Perova, N.B., Psovsky, Yu.P., Fedorovich, V.P. and Frolov, M.S. 1971, 1974, 1976. Supplements to the General Catalogue of Variable Stars, Sternberg State Institute, Moscow.

Lasker, B.M., and Hesser, J.B. 1969. Ap.J., 158, L 171.

Lasker, B.M., and Hesser, J.E. 1971. Ap.J., 163, L 89.

Ludendorff, H. 1922. A.N., 217, 167.

Luyten, W.J. 1949. Ap.J., 109, 528.

Luyten, W.J. and Hughes, H.S. 1965. Search for Faint Blue Stars, University of Minnesota, No. 36.

Mayall, M.W. 1949. Harvard Obs.Bul., 919.

Mayall, M.W. 1957, J.R.A.S.C., 51, 308.

Mumford, G.S. 1956. P.A.S.P., 68, 538.

Mumford, G.S. 1963. A.J., 68, 286.

Mumford, G.S. 1964. P.A.S.P., 76, 57.

Mumford, G.S. 1965. A.J., 70, 686.

Mumford, G.S. 1967a. P.A.S.P., 79, 283.

Mumford, G.S. 1967b. Ap.J., Supp., 15, 1.

Mumford, G.S. 1969. Kitt Peak Contr. 405.

Mumford, G.S. 1970. Ap.J., 162, 363.

Mumford, G.S. 1971a. Ap.J., 165, 369.

Mumford, G.S. 1971b. P.A.S.P., 80, 175.

Mumford, G.S. 1976. I.B.V.S., 1109, 1133.

Mumford, G.S., and Krzeminski, W. 1969. Ap.J., Supp., 18, 429.

Münch, G., and Morgan, W.W. 1953. Bol. Ton. y Tac., 8, 22.

Mustel, E.R., and Antipova, L.I. 1971. Nauch.Inf., No. 19, 32.

Nather, R.E., and Robinson, E.L. 1974. Ap.J., 190, 638.

Nather, R.E., and Warner, B. 1969. M.N.R.A.S., 143, 145.

Paczynski, B. 1965a. A.A., 15, 305.

Paczynski, B. 1965b. A.A., 15, 197.

Payne-Gaposchkin, C. 1969. Ap.J., 158, 429.

Peek, B.M. 1924. M.N.R.A.S., 84, 541.

Peek, B.M. 1925. M.N.R.A.S., 85, 666.

Peek, B.M. 1926. M.N.R.A.S., 86, 370.

Peek, B.M. 1927. M.N.R.A.S., 87, 573.

Peek, B.M. 1928. M.N.R.A.S., 88, 704.

Peek, B.M. 1931. M.N.R.A.S., 91, 203.

Plavec, M., and Polidan, R.S. 1975. Nature, 253, 173.

Pottasch, S. 1959. An d'Ap., 22, 412.

Pottasch, S. 1967. B.A.N., 19, 227.

Richer, H.B., Auman, B.C., Isherwood, B.C., Steele, J.P., and Ulrych, T.J. 1973. Ap.J., 180, 107.

Robinson, E.L. 1973a. Ap.J., 181, 531.

Robinson, E.L. 1973b. Ap.J., 183, 193.

Robinson, E.L. 1973c. Ap.J., 186, 347.

Robinson, E.L. 1974. Ap.J., 193, 191.

Robinson, E.L. 1975. A.J., 80, 515.

Robinson, E.L. 1976. In press.

Robinson, E.L., Nather, R.E., and Kiplinger, A. 1974. P.A.S.P., 86, 401.

Robinson, E.L. and Warner, B. 1972. M.N.R.A.S., 157, 85.

Rosino, L. and Tempesti, P. 1976. Soviet Astronomy, in press.

Sawyer Hogg, H., and Wehlau, V. 1964. A.J., 69, 141; 1965. A.J., 70, 678.

Schild, R. 1969. Ap.J., 157, 708.

Semeniuk, I. 1975. I.B.V.S., No. 1058.

Semeniuk, I., Kruszewski, A., and Schwarzenberg-Czerny, A. 1976. I.B.V.S., 1157.

Shugarov, S.Yu. 1975. A.C., No. 887.

Smak, J. 1962. A.J., 67, 643.

Smak, J. 1969. A.A., 19, 287.

Smak, J., and Stepien, K. 1969. Non-periodic Phenomena, 335.

Steavenson, W.H. 1921. M.N.R.A.S., 82, 403.

Steavenson, W.H. 1923. M.N.R.A.S., 83, 397.

Steavenson, W.H. 1924. M.N.R.A.S., 84, 538.

Steavenson, W.H. 1925. M.N.R.A.S., 85, 665.

Steavenson, W.H. 1926. M.N.R.A.S., 86, 365.

Steavenson, W.H. 1927. M.N.R.A.S., 87, 568.

Steavenson, W.H. 1928. M.N.R.A.S., 88, 615.

Steavenson, W.H. 1929. M.N.R.A.S., 89, 697.

Steavenson, W.H. 1930. M.N.R.A.S., 90, 705.

Steavenson, W.H. 1931. M.N.R.A.S., 91, 941.

Stephenson, B., Sandage, A., and Schild, R. 1968. Ap.Letters, 1, 247.

Strohmeier, W. 1963. I.B.V.S., 47.

Szkody, P. 1974a. P.A.S.P., 86, 38.

Szkody, P. 1974b. Ap.J., 192, L 75.

Szumiejko, E. 1976. This meeting.

Tempesti, P. 1975a. I.B.V.A., 974.

Tempesti, P. 1975b. IAU Card 2834.

Tempesti, P. 1976. I.B.V.S., 1098.

Thackeray, A.D. 1959. M.N.R.A.S., 119, 629.

Thackeray, A.D., and Hutchings, J.B. 1974. M.N.R.A.S., 167, 319.

Thomas, H.L. 1940. Harvard Obs. Bul., 912.

Vogt, N. 1974. Astr. and Ap., 36, 369.

Vogt, N. 1975. Astr. and Ap., 41, 15.

Wallerstein, G. 1963. P.A.S.P., 75, a6.

Wallerstein, G. 1969. P.A.S.P., 81, 672.

Walker, M.F. 1954. P.A.S.P., 66, 71.

Walker, M.F. 1956. Ap.J., 123, 68.

Walker, M.F. 1961. Ap.J., 134, 171.

Walker, M.F. 1963a. Ap.J., 137, 485.

Walker, M.F. 1963b. Ap.J., 138, 3 3.

Walker, M.F. 1965. Sky and Tel., 29, 23.

Walker, M.F. 1966. Advances in Elec. and Phys., 228, 761.

Walker, M.F., and Chincarini, G. 1968. Ap.J., 154, 157.
Walker, M.F., and Herbig, G.H. 1954. Ap.J., 120, 278.
Walker, M.F., and Reagan, G.H. 1971. I.B.V.S., 544.
Warner, B. 1971. P.A.S.P., 83, 817.
Warner, B. 1972a. M.N.R.A.S., 158, 425.
Warner, B. 1972b. M.N.R.A.S., 159, 95.
Warner, B. 1972c. M.N.R.A.S., 159, 315.
Warner, B. 1973a. M.N.R.A.S., 163, 25 P.
Warner, B. 1973b. M.N.A.S.S.A., 32, 120.
Warner, B. 1974. M.N.R.A.S., 168, 235.
Warner, B. 1975a. M.N.R.A.S., 170, 219.
Warner, B. 1975b. M.N.R.A.S., 173, 37 P.
Warner, B. and Brickhill, A.J. 1974. M.N.R.A.S., 166, 673.
Warner, B. and van Citters, G.M. 1974. Obs., 94, 116.
Warner, B. and Harwood, J.M. 1973. I.B.V.S., 756.
Warner, B. and Nather, R.E. 1971. M.N.R.A.S., 152, 219.
Warner, B. and Nather, R.E. 1972a. M.N.R.A.S., 156, 1.
Warner, B. and Nather, R.E. 1972b. M.N.R.A.S., 156, 297.
Warner, B. and Nather, R.E. 1972c. M.N.R.A.S., 156, 305.
Warner, B. and Robinson E.L. 1972. M.N.R.A.S., 159, 101.
Warner, B. and Thackeray, A.D. 1975. M.N.R.A.S., 172, 433.
Weber, R. 1960. J.O., 44, 275.
Wegner, G. 1972. Astrophys.Letters, 12, 219.
Widing, K.G. 1966. Ap.J., 143, 121.
Wright, K.O. 1957. P.A.S.P., 69, 522.

OBSERVATIONS OF NOVAE AND RELATED OBJECTS AT MINIMUM LIGHT

Brian Warner

Department of Astronomy, University of Cape Town

ABSTRACT

Mass determinations indicate that the white dwarf components of cataclysmic variables are heavier than $0.5M_\odot$; this is in harmony with the themonuclear theories of novae explosions which demand masses of at least $0.5M_\odot$.

Absolute magnitudes of dwarf novae are $M_V \sim 7.5$ and of classical novae $M_V \sim 4.5$. This difference is not understood, but appears to be associated with the accretion disc luminosities rather than the properties of the stellar components. In some cases the accretion discs are so bright that a doppler-broadened absorption spectrum can be seen.

The rate of mass transfer may account for the different appearances of dwarf and classical novae. There is still no satisfactory theory for the driving mechanism of the mass transfer process.

A full review of these and other aspects of cataclysmic variables is given in the author's review "Observations of Dwarf Novae" appearing in Structure and Evolution of Close Binary Systems (I.A.U. Symposium No. 73).

M. Friedjung (ed.), Novae and Related Stars, 33. *All Rights Reserved.*
Copyright © 1977 by D. Reidel Publishing Company, Dordrecht, Holland.

MASS TRANSFER AND ACCRETION DISC FLOW IN CLOSE BINARY SYSTEMS

D.N.C. Lin and J.E. Pringle

Institute of Astronomy, Cambridge.

INTRODUCTION

Recent review articles by Warner (1976, 1977) and by Robinson (1976) summarize our present understanding of the structure of the cataclysmic variable binary systems, and of dwarf novae in particular. In this contribution we turn our attention to an understanding of the gas flows associated with the mass transfer process and with the setting up of an accretion disc around the white dwarf (primary). Close to the primary the gas flow can be modelled using standard accretion disc theory (Prendergast and Burbidge 1968, Bath, Evans, Papaloizou and Pringle 1974) except at the surface of the white dwarf where emission from the boundary layer should be considered (Lynden-Bell and Pringle 1974, Pringle 1977). Most of the optical emission in dwarf novae, however, comes from much larger radii, where the temperatures are lower, and unfortunately where the structure of the gas flow is much more complicated. The interpretation of optical observations is expected to be further confused by a dominant part of the optical emission from the regions being due to reprocessed ultraviolet and soft X-ray flux originating close to the primary. In particular gas motion in the direction perpendicular to the orbital plane - usually ignored in model gas flow calculations - can strongly affect the orbital light curve. We divide our discussion into two parts. First we review the results obtained from numerical modelling of the gas flow. Second we consider the ways in which optical observations of the systems can be used to test these results.

M. Friedjung (ed.), Novae and Related Stars, 35-40. All Rights Reserved.
Copyright © 1977 by D. Reidel Publishing Company, Dordrecht, Holland.

OBSERVATIONAL DISCS

An estimate of the mass of a dwarf nova accretion disc can
be made as follows. The rates of change of orbital period (or
upper limits thereto) interpreted naively as mass transfer rates
give values $\leq 3 \times 10^{-7}$ $M_\odot$ y^{-1} (Pringle 1975). If the outbursts of
dwarf novae are accepted as accretion events, then the decay
timescale of the outburst light curve (a few days) corresponds to
the viscous timescale in the disc. Thus if matter flows continuously through the disc, the average mass of gas inside the
primary's Roche lobe is expected to be $\sim 10^{-9}$ $M_\odot$. On the other
hand if the transferred matter is accumulated in the lobe between
outbursts, which occur, say, every ~ 100 days, then the maximum
mass of gas in the disc is $\sim 10^{-7}$ $M_\odot$. For the reasons discussed
by Pringle (1975) the estimates of the mass transfer rates, and
hence of the disc masses, are likely to be overestimates. There
are two further observational estimates of disc masses in the
literature. First an estimate of $\sim 10^{-5}$ $M_\odot$ is made by Smak (1972)
in order to explain a supposed quasi-sinusoidal variation in the
O-C curve of U Gem with a ~ 10 year timescale in terms of angular
momentum exchange between the disc and the orbit. It is not
clear to us how significant the O-C variation is. Even if the
variation is confirmed, the explanation given by Smak is certainly
not unique. For example the ~ 30 year variation in the O-C curve
of UX Ursae Majoris has been explained both in terms of a third
body in the system (Nather and Robinson 1974) and in terms of
apsidal motion (Africano and Wilson 1976). Second an estimate of
$\sim 10^{-5}$ $M_\odot$ is made by Warner (1974) in order that he may interpret
the few day decay timescale of the outburst of the dwarf nova
Z Cha as the cooling timescale of the disc. Such an interpretation has been questioned by Bath et al (1974).

There are a number of ways in which the physical size of the
disc in dwarf nova systems can be estimated observationally. The
most direct photometric method is that employed by Warner (1974)
who observed the eclipses of Z Cha during outburst and during
quiescence to show that the disc, and not just the primary, is
involved in the optical outburst. Optical photometric observations are somewhat limited, however, since the disc is not expected to be uniformly bright at optical wavelengths. For example,
since for a steady disc emitting locally as a black body the
temperature $T \propto R^{-\frac{3}{4}}$, the inner regions where $T \gtrsim 10^4$ K, though
dominating the bolometric luminosity, would appear relatively
unluminous to an optical observer. Reflection effects exaggerate
this conclusion. The outer parts of the disc may be optically
thin and radiate mostly by line emission (Robinson 1976). In
this case, to estimate the outer radius of the disc, timedependent
spectroscopic observations may be more appropriate.

The "hump" in dwarf nova light curves - typified by the
light curve of U Gem (Krzeminski 1965) - can give some insight
into the gas flow. In particular, the use of the hump in the
interpretation of the light curve of U Gem by Smak (1971) and by
Warner and Nather (1971) provides the basis for our present understanding of these systems. The hump is thought to be caused by
the anisotropy of the radiating region associated with the "hot
spot" where the stream of transferred material strikes the
material already circling the primary. Detailed conclusions
drawn from the shape and size of the hump in the light curve,
especially about the position of the disc/stream interaction and
about the mass transfer rate should be treated with care, for a
number of reasons. First, since the flow timescales in the disc
are much less than the orbital timescales, apparent humps can
occur in the light curve due to an overall luminosity variation
rather than to the presence of a hot spot. This is particularly
true during outburst. Second, the radiative processes occurring
in the hot spot are not understood. The energy dissipated in
the interaction can be carried downstream before being radiated.
The hump is only a measure of the anisotropic part of the flux
radiated at the hot spot. The optical flux emitted by the hot
spot is not the total flux.

Spectroscopic observations around the binary orbit, especially during the primary eclipse, should provide insight into the
conditions in the disc and associated gas streams. In addition
the behaviour of the shape of the lines through eclipse can provide clues to the overall size of the disc. The breadth of
lines, giving a measure of circular velocities in the disc, can
indicate how far inwards the disc (or rather the line producing
part of the disc) extends (Warner 1976).

THEORETICAL DISCS

It is usually assumed that the stream of transferred material
in dwarf novae does not strike the white dwarf primary directly.
For the systems whose parameters are known, this assumption is
valid provided that the mass losing secondary corotates with the
binary system and that the initial velocity of the transferred
material is not supersonic with respect to the sound speed in the
secondary's atmosphere. When mass transfer commences, the stream
of material circles the primary, following a particle trajectory,
until it intersects itself. Most of the interacting gas is now
sufficiently deep in the potential well of the primary that during
the resulting rapid dissipation angular momentum about the primary
is approximately conserved. We may therefore expect the material
to form a ring around the primary with radius R_h such that the
specific angular momentum of the stream about the primary before
the dissipation is roughly equal to $(GM_1R_h)^{\frac{1}{2}}$. This circularization

process occurs on a timescale $\sim(R_h^3/GM_1)^{\frac{1}{2}}$, corresponding to the
orbital period about the primary at that radius (cf Webbink 1976).
This is in general much shorter than the binary period. If there
are no processes present locally in the resulting gas flow which
can transport momentum or angular momentum (we shall refer to
such processes as viscous processes) then the transferred
material continues to accumulate in a ring at radius $\sim R_h$. In
particular, no material can be accreted by the primary.

Numerical calculations of the process just described have
been carried out by Flannery (1975). He uses an explicit itera-
tive scheme to model the gas flow, with no additional viscous
terms, and continues the integration for one binary period after
the onset of mass transfer. His calculations confirm the expected results described above, except that the velocities of the
gas in the ring appear to have additional <u>outward</u> radial components. This may be a transient effect or it may be due to the
explicit nature of his iteration scheme. Prendergast and Taam
(1974) have performed numerical calculations for the case when
the primary is large enough that the stream of transferred
material strikes it directly. These calculations are therefore
not directly relevant to dwarf novae but might provide a reasonable estimate of the gas flow when the disc around the primary is
much denser than the stream of incoming material. Even in this
case, however, their results should be treated with caution since
the numerical code used does not explicitly conserve angular
momentum (with respect to an inertial frame).

Most authors appear to agree that the white dwarf in dwarf
nova binaries accretes material for at least some of the time.
This implies that viscous processes must occur in the fluid flow
at some stage, which in turn implies that the disc is expected to
be larger than our initial estimate of $\sim R_h$ (Lynden-Bell and
Pringle 1974, Lin and Pringle 1976). This is simply because in
order that some matter lose enough angular momentum with respect
to the primary to be accreted, other matter must gain some and so
move to radii larger than R_h. Note, however, that at radii $>>R_h$
and comparable with the Roche lobe radius R_L, the concept of
angular momentum about the primary is no longer particularly
useful. Although the gas flow problem in which the mass transfer
rate is timedependent and in which the viscosity depends generally
upon conditions in the flow is likely to remain intractable for a
while, the problem in which the mass transfer rate is constant
and the viscosity is a given function of position is more approach-
able and should give some insight into the more general problems.
A first attempt at the simpler problem has been made by Lin and
Pringle (1976), with a view to answering two specific but inter-
related problems: (i) how large is the accretion disc around the
primary likely to be and (ii) how much of the material transferred
through the inner Lagrangian point is actually accreted by the

primary. There are two obvious ways in which the excess angular
momentum can be disposed of. First, material can flow out of the
primary's Roche lobe carrying the excess angular momentum with it.
In this case we may expect the accretion disc to be comparable in
size to the Roche lobe and the fraction of transferred matter
that is not accreted to be $f \sim (R_h/R_L)^{\frac{1}{2}} \sim 30\%$ (Prendergast and
Burbidge 1968). Second, tidal interactions between the disc and
the secondary can transfer angular momentum from the disc into
orbital motion (Börner et al 1973). In this case there is little
need for the disc to be as large as R_L. In addition almost all
the transferred material could in principle be accreted by the
primary. Note that both these processes depend on the existence
of a viscosity. The ratio of the effectiveness of tidal to mass
flow transport is likely to be relatively independent of the
viscosity. It is a steeply increasing function of the radius of
the disc. Lin and Pringle model the fluid flow by means of time
averaging the motions of a large number (a few thousand) of
"viscously interacting" particles. The reader is referred to Lin
and Pringle (1976) for further details. They find that the size
of the disc is comparable to that of the primary's Roche lobe but
that only a few per cent ($f \sim 1-3\%$) of the transferred material is
not accreted by the primary. This small figure (compared to
$\sim (R_h/R_L)^{\frac{1}{2}}$) is a measure of the effectiveness of tidal processes.
Of course the disc should, in a steady state, be expected to
extend to the Roche lobe since the effectiveness ratio for the
two processes does not become infinite at any radius. The above
calculations were carried out for a corotating secondary and a
low velocity through the inner Lagrangian point. Further calculations, relaxing these assumptions, are in progress.

CONCLUSION

     In summary, it seems that dwarf novae and related systems
can provide us with much-needed information on the mechanism of
mass transfer in close binary systems. Numerical simulations of
the gas flow to date leave a lot to be desired but do give some
insight into the processes occurring. Optical observations with
high speed photometric and spectrophotometric systems are especially valuable, although it should be borne in mind that the
optical emission from dwarf novae is probably a small fraction of
their total luminosity.

REFERENCES

Africano, J., and Wilson, J., 1976. Pub.astron.Soc.Pacific., 88,
     8.
Bath, G.T., Evans, W.D., Papaloizou, J., and Pringle, J.E., 1974.
     Mon.Not.R.astr.Soc., 169, 447.

Börner, G., Meyer, F., Schmidt, H.U., and Thomas, H.-C., 1973.
    Mitt.der.Astr.Ges., 32, 237.
Flannery, B.P., 1975. Astrophys.J., 201, 661.
Krzeminski, V., 1965. Astrophys.J., 142, 1051.
Lin, D.N.C., and Pringle, J.E., 1976. Proc. I.A.U. Symp. 73.
    eds P.P. Eggleton et al, Reidel, Dortrecht, in press.
Lynden-Bell, D., and Pringle, J.E., 1974. Mon.Not.R.astr.Soc.,
    168, 603.
Nather, R.E., and Robinson, E.L., 1974. Astrophys.J., 190, 637.
Prendergast, K.H., and Burbidge, G.R., 1968. Astrophys.J.Lett.,
    151, L83.
Prendergast, K.H., and Taam, R.E., 1974. Astrophys.J., 189, 125.
Pringle, J.E., 1975. Mon.Not.R.astr.Soc., 170, 633.
Pringle, J.E., 1977. Mon.Not.R.astr.Soc., in press.
Robinson, E.L., 1976. Ann.Rev.Astr.Astrophys., in press.
Smak, J., 1971. Acta Astronomica, 21, 15.
Smak, J., 1972. Acta Astronomica, 22, 1.
Warner, B., 1974. Mon.Not.R.astr.Soc., 168, 235.
Warner, B., 1976. Proc. I.A.U. Symp. 73. Eds P.P. Eggleton et al,
    Reidel, Dortrecht, in press.
Warner, B., 1977, this volume.
Warner, B., and Nather, R.E., 1971. Mon.Not.R.astr.Soc., 152, 219.
Webbink, R.F., 1976. Nature, 262, 271.

TRANSFER INSTABILITIES AND ACCRETION DISC STRUCTURE IN DWARF NOVAE

G.T. Bath

Dept. of Astrophysics, Oxford.

INTRODUCTION

In the past four years the importance of accretion in dwarf nova binaries has been increasingly appreciated. It is now widely accepted that their eruptive activity is most probably due to bursts of accretion flow through an accretion disc onto the blue component. Two possible mechanisms for modulating the accretion process into quasi-periodic bursts have been suggested. Either the rate of mass transfer by the red component is modulated by enhanced outflow during each eruption (Bath 1973, 1975; Gorbatskii 1975; Papaloizou and Bath 1975), or, alternatively, the disc itself (continuously accreting from the red component in a steady way) undergoes some form of intrinsic instability which results in the infall of material normally stored in the outer disc regions (Osaki 1974). In this paper some of the general arguments which support these accretion models will be outlined, and the present status of the two accretion theories reviewed. A fuller discussion will be found in Bath (1976). The main purpose of this paper is, firstly, to emphasise the importance of the observational relationship found by Bailey (1975) between the outburst decline rate and the binary period, and, secondly, to note the uncertainty in the present estimates of the masses of accretion discs in dwarf novae.

GENERAL ARGUMENTS FAVOURING ACCRETION

It is now well established (Smak 1971b; Warner 1974a) that the outbursts of dwarf novae occur in the accretion disc surrounding the blue component. Since an accretion disc necessarily radiates

M. Friedjung (ed.), Novae and Related Stars, 41-49 All Rights Reserved.
Copyright © 1977 by D. Reidel Publishing Company, Dordrecht, Holland.

as a result of viscous dissipation (thereby allowing material to
spiral in and be accreted by the white dwarf (Prendergast and
Burbidge 1968; Pringle and Rees 1972; Shakura and Sunyaev 1973;
Rees 1976), it is evident that such disc brightening could be
produced by enhanced mass transfer and enhanced dissipation in the
disc.

The luminosity, L_a, is given by the rate at which the
potential energy of infall of the accreted material is liberated,

$$L_a = \frac{GM_1}{R_1} \; \dot{m} \quad \text{erg s}^{-1}$$

Thus $L_a \sim 10^{17} \dot{m}$ for values of M_1 and R_1 relevant to white
dwarfs. The observed luminosity of dwarf novae in their
quiescent state $\sim 10^{32}$ ergs^{-1} (Kraft 1962; Kraft and Luyten
1965) and at outburst $\sim 10^{34} - 10^{35}$ erg s^{-1} (Warner 1974b).
Thus the quiescent luminosities correspond to mass transfer rates,
$\dot{m} \sim 10^{15}$ gm s^{-1} ($10^{-11} M_\odot$ yr^{-1}) and the outbursts to
$\dot{m} \sim 10^{17} - 10^{18}$ gm s^{-1} ($10^{-9} - 10^{-8} M_\odot$ yr^{-1}).

The resulting optical luminosity depends on the bolometric
corrections and therefore on the temperature. The characteristic
temperature, T_d, of the inner disc region, where most of the
accretion luminosity is radiated, is approximately the black body
temperature of a source of dimensions $\sim R_1$ and luminosity $\sim L_a$,

$$T_d \sim \left(\frac{L_a}{4\pi\sigma R_1^2}\right)^{1/4} \sim \left(\frac{GM_1 \dot{m}}{4\pi\sigma R_1^3}\right)^{1/4} \quad {}^\circ K$$

$T_d \sim 2 \times 10^4 \; {}^\circ K$ in the quiescent state and $T_d \sim 7 \times 10^4 -$
$2 \times 10^5 \; {}^\circ K$ in the outburst phase. Thus in the quiescent state
the disc will radiate predominantly in the optical (mainly in the
lines in the outer disc regions where $T_d \lesssim 10^4 \; {}^\circ K$). At outburst
the source will radiate strongly in the ultraviolet, possibly
extending to the soft x-ray region.

All these features are consistent with the observed
behaviour. Detailed disc models confirm these conclusions, with
minor changes in the quantitative details (Bath et.al 1974;
Evans 1974; Bath 1976; Pringle 1976; Tylenda 1977). Note
that the mass transfer rates, even in the outburst phase, are
quite moderate. They allow lifetimes $\sim 10^9$ yr. The rate of
mass transfer through the disc in the quiescent state is at least
a factor 10^{-2} less than the rate at outburst (and could be even
less if the intrinsic luminosity of the white dwarf dominates at

quiescence. The brightness of the hot-spot shoulder makes this unlikely, however.)

If the outbursts are generated by instabilities in the red component then the quiescent mass transfer rate is sufficiently small that it should not radically affect the envelope structure of the star. This quiescent mass loss may be regarded as a wind generated above a stellar envelope which is close to hydrostatic equilibrium. The values of $\dot{m}$ at outburst, on the other hand, are precisely those which would be driven by dynamical instabilities in which the acceleration terms are important. Such instabilities have been demonstrated to exist, and to have the required recurrence times (Papaloizou and Bath 1975; Bath 1975). According to this work the red component envelope becomes dynamically unstable when the photosphere is at, or close to, the critical inner Lagrangian point. The instability arises as a result of the reduced gravity in this region and the destabilising effects of sub-photospheric ionization zones, and associated super-adiabatic gradients. The ensuing outburst drives the envelope progressively away from thermal equilibrium. Eventually this increasing disequilibrium damps the instability, the envelope collapses, and slowly recovers thermally through energy transport from the interior. In this quiescent phase the photosphere slowly expands, eventually regaining the initial equilibrium state and undergoing a second dynamical instability. The outburst period is determined by the thermal relaxation time for the envelope, given approximately by

$$\tau = \frac{R\,\bar{T}\,\Delta M}{\Delta L}$$

where $\bar{T}$ is the mean temperature of the disturbed outer envelope of mass ΔM suffering a luminosity deficit ΔL.

The process is a relaxation phenomenon in the sense that the greater the mass lost each outburst, the greater ΔM and $\bar{T}$, and the longer the recovery time to the next outburst. Greater mass loss leads to a greater release of total energy through accretion onto the white dwarf. Thus large outbursts are associated with long recovery periods, and vice versa. This is exactly the sort of relaxation phenomenon observed in the eruptions of dwarf novae (Kruytbosch 1928; Sterne and Campbell 1934) and is illustrated by the well-known Kukarin-Parengeno relationship.

It has been objected (Warner and Nather 1971) that Krezminski's (1965) observations of the hot-spot shoulder and eclipse in U Gem precludes unstable mass transfer originating in the red component as the cause of accretion modulation in the disc. Since the spots optical luminosity appears to change by a

factor $\lesssim$ 4 it is concluded that the mass transfer rate through
the spot cannot have changed. However there is strong evidence
that in many other systems the hot-spot shoulder suffers large
changes in both binary phase and amplitude (Smak 1971a; Warner
1974a).   Such changes in phase are most naturally accounted for
not by changes in the actual position of the spot on the edge of
the disc, but by variations in its radiation pattern with respect
to binary phase. Thus if the spot were to start to radiate more
strongly in a direction towards the blue component as a result of
optical depth effects, the hot spot shoulder would appear to
change in phase through half the binary orbit.   Naively
interpreted it would appear to have moved in position around the
disc.   Similarly, the amplitude of the shoulder, at whatever
phase it appears, is determined by the anisotropic part only of
the radiation field of the spot (Pringle 1976).   Changes in
shoulder amplitude are not necessarily due to changes in the spot's
total intrinsic luminosity and vice versa.   The variety of
behaviour exhibited by the shoulder and hot-spot in different
systems - the very small changes in U Gem (Smak 1971b), the
luminosity changes expected for unstable mass transfer in Z Cha
(Bath et. al. 1974), the phase variations observed in many
systems, and Z Cha in particular (Smak 1971a;Warner 1974a), the
change in binary period of $\sim$ 3% exhibited by the shoulder in the
outburst of VW Hyi (Vogt 1974; Warner 1975) - illustrate the
difficulties which attend any attempt to use the binary light
curve shoulder as a simple diagnostic tool.

It should be noted, furthermore, that the implicit assumption
that the intrinsic optical luminosity of the spot necessarily
reflects the rate of mass flow through it in non-steady conditions
has not been justified by any detailed theoretical model.   In the
case of steady conditions (Warner and Nather 1971;   Smak 1971b;
Lubow and Shu 1975;   Bath et.al. 1974) it can be crudely justified,
but sudden outflows of short duration by the red star may not allow
the spot sufficient time to relax, and reflect this sudden dis-
charge by changes in luminosity.   Note also that the outburst
time of the red component is typically 1-12 hrs (Bath 1975), which
is of the order of the binary period itself, and not necessarily
easily observed. The decay time of the eruption (except in
systems exhibiting Z Cam standstills) is determined by the timescale for viscous dissipation in the disc (Bath et.al. 1974), not
by the timescale for the decay of the outburst in the red
component (see also section C).   Note that at mass transfer
rates $\dot{m} \sim 10^{17} - 10^{18}$ gm s^{-1}, outbursts lasting 1-12 hr transfer$\sim$
$10^{21} - 10^{23}$ gm ($10^{-12} - 10^{-10} M_\odot$).   As pointed out earlier the
energy per gram produced by accretion, $GM_1/R_1 \sim 10^{17}$ erg gm^{-1}.
The predicted mass per burst of $10^{21} - 10^{23}$ gm therefore generates
$10^{38} - 10^{40}$ erg, which is the same as the observed energy released
per eruption.

The alternative accretion model suggested by Osaki (1974) remains largely unexplored, at least in terms of detailed theoretical models.[1] Whether an instability of the required nature exists which possesses the correct recurrent times and which operates at the required mass transfer rates is not yet known. Further understanding of disc structure, and of possible sources of instability, are required before a full investigation of this suggestion can be undertaken. Its chief attraction is that it allows, in a simple way, the hot-spot in U Gem to remain basically unaffected by the eruption. But then the question immediately arises - why is the spot affected so markedly in other systems? And is it purely fortuitous that the properties of stellar envelope instabilities fit the outburst properties, not only of dwarf novae, but also of symbiotic variables, despite their large differences in structure? And is it only coincidence that Webbink's (1976) similar model of the recurrent nova TcrB accounts for a number of unexplained phenomena in its outbursts?

THE MASS OF THE DISC

Present estimates of disc masses in dwarf novae are $\sim 10^{-5} M_\odot$ (Smak 1972); Warner 1974a). Such large values are not consistent with the same disc viscosity as that operating in x-ray sources. If the same viscosity were acting in both cases then the mass of the disc should be $\sim 10^{-10} M_\odot$ (Lin & Pringle 1977). Only if the viscosity is very much less in dwarf novae could the mass be greater. In that case, of course, the accretion rates would be reduced, and could not account for the disc luminosity. We will examine first the basis for the earlier mass estimates of $10^{-5} M_\odot$, and then present evidence that suggests that the masses are $\sim 10^{-10} M_\odot$, and consistent with present disc accretion models.

The estimate of Warner (1974) is based on the assumption that there is no dissipation in the disc, no internal viscous energy sources, and hence no accretion. He identifies the thermal cooling time of this "disc" with the observed decline time, and deduces a mass of $3 \times 10^{-5} M_\odot$. He assumes the initial heating is due to a shock generated by explosive nuclear burning in the underlying white dwarf. There are two major objections to this approach. Firstly, it is not relevant to a dissipative disc model of the type considered in this article, in which the energy during the outburst is generated by accretion. Secondly theoretical work has shown that it is difficult, if not impossible, to propagate

[1] We note that Inoue (1976) has recently discussed the possibility of intermittent accretion due to Rayleigh-Taylor instabilities in a magnetised disc in which both this disc and accreting star have magnetic fields with parallel polarities.

a shock through a differentially rotating disc, due to the
decreased importance of pressure support (Sanders and Prendergast
(1974; Bath et.al 1974; Sparks 1976).

The estimate of Smak (1972) is based on a model for the
alternating binary period change observed in U Gem. It is
assumed that these period changes are due to angular momentum
storage in the disc. We first note that such storage occurs
in the outer envelope of the blue star, in standard accretion
disc models, and the limit of $10^{-4} \gtrsim M/M_\odot \gtrsim 10^{-6}$ may refer to the
mass of this accreted rotating envelope. Such a mass in the
envelope would be built up in $10^4 - 10^6$ years at the accretion
rates discussed in the previous section. Secondly Pringle (1975)
has questioned the reality of the period changes in this system.
Thirdly, many factors, such as tidal dissipation, mass and
angular momentum loss, non-synchronism of the binary components,
etc, can produce changes of period, and have not been excluded
as an alternative explanation.

Indirect evidence in favour of low disc masses is provided
by the outburst energetics. The accreted mass per outburst, ΔM,
is just

$$\Delta M = \frac{\Delta E \; R_1}{GM_1} \quad \text{gm}$$

where ΔE is the total energy per outburst.
For $\Delta E \sim 10^{39}$ erg, $\Delta M \sim 10^{-10} M_\odot$. Thus if the mass of the
disc were $10^{-5} M_\odot$ each outburst would correspond to a minute
fraction (10^{-5}) of the total disc mass being deposited on the
white dwarf surface. Each outburst would correspond to an
infinitesimal mass flow through an underlying massive disc, or
from a massive outer ring. Such behaviour is difficult to
conceive being possible according to present understanding of disc
structure. Furthermore, note that the mass deposited in each
burst from energetic considerations is the same as that of the
standard disc accretion models and compatible with present
viscosities.

The observed relationship found by Bailey (1975) between the
binary period and the outburst decline rate, may prove to be the
most important empirical relationship in this regard. Evidence
is presented in this paper which indicates that the longer period
systems suffer eruptions which decline on a longer timescale than
the shorter period systems. The relation between binary period
and decline rate is approximately linear. Although more extensive
investigation of this correlation is still required, such a
relationship has profound implications for models of the outbursts.

It indicates that the eruption light curves are controlled by the binary orbital parameters.   In particular, for a dissipative disc model, in which the light curves are generated by the slow diffusion of material through the disc, the decline rate is determined by the viscosity, ν , and the radius at which suddenly transfered material first goes into orbit, R_s.   If the matter is injected as a burst, then the decline rate is determined by these two factors, essentially by the ratio R_s^2/ν (Lynden-Bell and Pringle 1974; Bath et. al 1974).   It is easy to show that Bailey's relationship implies a viscosity which is the same in all systems, and the same order as that occuring in x-ray sources. Further work on this empirical relationship could provide important information on disc structure and outburst behaviour in these systems.

SUMMARY AND CONCLUSION

    The realisation that the processes occuring in dwarf novae are essentially similar to those occuring in x-ray sources, with the substitution of a white dwarf for a neutron star as the accreting primary, has opened up a whole new perspective on these systems.   The essential features seem well accounted for by accretion models.   From a theoretical point of view the main question remaining is the cause of unstable mass flow through disc during eruptions.   Both episodes of enhanced mass transfer by the red component and quasi-periodic instabilities in the disc have been advanced.   In the former case detailed models give a good fit to the observations.   Much work remains to be done in the latter case.   Obvious areas for future research are:

    A) Red component structure and mass transfer behaviour. Further work on the linear stability of stars conforming to the Roche potential, extending that of Papaloizou and Bath (1975), would clearly be worth while.   In the case of non-linear models, the inclusion of time-dependent convection (if only schematically) is a priority.   Ideally a two dimensional hydrodynamic non-linear study is required.

    B) Disc structure, with emphasis on time-dependent behaviour and possible sources of instability.   Priorities include both theoretical investigation of the microphysics of the disc - origin of viscosity, local and global stability, time-dependent behaviour, disc coronae/chromospheres, etc. - extending the work of Lynden-Bell, Prendergast, Sunyaev, Shakura, Pringle, Rees and others - together with a more basic exploration of disc properties deduced directly from the observed behaviour of dwarf nova outbursts.   The latter, extending the work of Smak and Warner, must include, in particular, further discussion of estimated masses and radii, and changes in structure that occur during outburst.   The relation between the outburst decline rate and binary period found by Bailey could provide much needed information on the magnitude and source of viscosity in discs in dwarf novae.

   C) Hot spot structure and the radiation pattern generated in the spot region as a result of both steady and unsteady mass transfer.   The two questions, namely what is the radiation pattern of the spot, and how is it affected by sudden "sneezes" in the mass transfer rate, are of primary importance in interpretation of the observational behaviour of the binary light curve shoulders.

   Finally, two observed phenomena are still fundamentally unresolved theoretically - the origin of Z Cam standstills and the V W Hydri supermaxima.   The former are almost certainly due to the systems occasionally getting stuck in a state of steady accretion, thus mimicing for a short time the state of old novae and nova-like variables.   Such a state could possibly be set up by the red component getting stuck in a quasi-steady outflow state with a more stable envelope structure.   The latter is clearly a case of major outbursts involving greater mass transfer per eruption.   These could perhaps be accounted for, at least in principle, by the interaction of the H/HeI ionization zone and the deeper HeII zone in the envelope of the red component.   The former dominate in normal destabilization of the envelope, but the latter could destabilize and produce occasional larger outbursts.   These would relax on a thermal timescale which was longer and produce occasional supermaxima with a longer mean period.

REFERENCES

Bailey, J., 1975, J.Brit.astr.Ass., 86, 30.
Bath, G.T., 1973, Nature Phys.Sci. 246, 84.
Bath, G.T., 1975, Mon.Not.R.astr.Soc. 171, 311.
Bath, G.T., 1976, In 'Structure and evolution of close binary
   systems', IAU Symp.73, D. Reidel, Dordrecht, Netherlands.
Bath, G.T., Evans, W.D., Papaloizou, J.C.B. & Pringle, J.E., 1974,
   Mon.Not.R.astr.Soc. 169, 447.
Evans, W.D., 1974, Unpublished thesis, University of Oxford.
Gorbatskii, V.G., 1975, Soviet Astronomy Letters, 1, 11.
Inoue, H., 1976, Publ.Astron.Soc.Japan 28, 293.
Kraft, R.P., 1962, Astrophys.J., 135, 408.
Kraft, R.P., & Luyten, W.J., 1965, Astrophys.J., 142, 1041.
Krezminski, W., 1965, Astrophys.J., 142, 1051.
Kruytbosch, W.E., 1928, B.A.N. 144, 145.
Lin, D.N.C., & Pringle, J.E., 1977, In this volume.
Lubow, S.H., & Shu, F.H., 1975, Astrophys.J., 198, 383.
Lynden-Bell, D. & Pringle, J.E., 1974, Mon.Not.R.astr.Soc., 168, 603.
Osaki, Y., 1974, Publ.Astron.Soc.Japan, 26, 429.

Papaloizou, J.C.B. & Bath, G.T., 1975, Mon.Not.R.astr.Soc.,
    172, 339.
Prendergast, K.H., & Burbidge, G.R., 1968, Astrophys.J.Lett.
    151, L83.
Pringle, J.E., 1975, Mon.Not.R.astr.Soc., 170, 633.
Pringle, J.E., 1976, p.191, In 'Structure and evolution of
    close binary systems', IAU Symposium 73, D. Reidel, Dordrecht,
    Netherlands.
Pringle, J.E., 1976, Mon.Not.R.astr.Soc. In press.
Pringle, J.E. & Rees, M.J., 1972, Astr.Astrophys. 21, 1.
Rees, M.J., 1976. In 'Structure and evolution of close binary
    systems', IAU Symposium 73, D. Reidel, Dordrecht, Netherlands.
Sanders, R.H. & Prendergast, K.H., 1974, Astrophys.J. 188, 489.
Shakura, N.J. & Sunyaev, R.A., 1973, Astr.Astrophys. 24, 337.
Smak, J., 1971a, IAU Colloquium 15, 248.
Smak, J., 1971b, Acta Astronomica 21, 15.
Smak, J., 1972, Acta Astronomica, 22, 1.
Sparks, W.M., 1976, Paper presented at the Illinois conference on
    Novae, Illinois.
Sterne, T.E. & Campbell, L., 1934, Ann.Harvard Coll.Obs., 90, 189.
Tylenda, R., 1977, In this volume.
Vogt, N., 1974, Astr. & Astrophys. 36, 369.
Warner, B., 1974a, Mon.Not.R.astr.Soc. South Africa, 33, 21.
Warner, B., 1975, Mon.Not.R.astr.Soc. 170, 219.
Warner, B. & Nather, R.E., 1971, Mon.Not.R.astr.Soc. 152, 219.
Webbink, R.F., 1976, Nature, 262, 271.

ON THE STABILITY OF ACCRETION DISKS

MARIO LIVIO and GIORA SHAVIV

DEPARTMENT OF PHYSICS AND ASTRONOMY
TEL-AVIV UNIVERSITY, RAMAT AVIV, ISRAEL

The steady state equilibrium equations for accretion disks are analysed for stability against pressure equilibrated perturbations and stability conditions are derived in the linear approxmation.

The conditions include : (a) The classical Schwarzschild condition in the vertical direction. (b) A modified condition in the radial direction, and (c) A condition on the energy generation due to viscous forces, which, when applied to the " -model" of viscosity yields an upper bound on of the order of 0.1.

We derive an upper limit to the luminosity of an accretion disk which is stable against convection, its value is quite below the Eddington limit.

M. Friedjung (ed.), Novae and Related Stars, 51. All Rights Reserved.
Copyright © 1977 by D. Reidel Publishing Company, Dordrecht, Holland.

RECENT PHOTOMETRIC OBSERVATIONS OF NOVA-LIKE VARIABLES

G. S. Mumford

Tufts University, Medford, Massachusetts

For a variety of reasons changes in the orbital periods of binary, nova-like variables are expected to occur. Photoelectric observations provide the most direct means of establishing such variations for those objects that have been found to eclipse. However, in the 15 to 20 years that have elapsed since periods for many were first determined, timings of additional minima have only been obtained sporadically. In many instances the coverage is neither complete nor up-to-date.

Data obtained in January, 1976, at Kitt Peak National Observatory confirmed the unchanging period of EX Hydrae and the currently increasing period of U Geminorum. Further observations are required to establish the true nature of the period variation of T Aurigae which appears, at present, to be cyclic.

From observations made in April, 1975, or even earlier, it has been tentatively concluded that the period of EM Cygni is decreasing, while that of DQ Herculis may be increasing. More data are clearly needed. In addition, as of the date above, the period of WZ Sagittae was unchanging. Insufficient data exist for V Sagittae so that not even a preliminary conclusion can be drawn.

CROSS CORRELATION DETECTION OF RAPID STELLAR VARIABILITY USING TWO TELESCOPES

P.J. Edwards & R.B. Hurst   M. Thomas

Physics Department   Beverly-Begg Observatory
University of Otago   Dunedin, New Zealand
Dunedin, New Zealand

The study of rapid, small amplitude stellar variability is handicapped by scintillation and photon noise. Flickering on time scales corresponding to bandwidths $B \gtrsim 0.1$ Hz may be expected of a variety of irregular variables including dwarf novae and X-ray binaries. Stochastic variability is harder to detect than periodic variability because its spectrum is likely to be a continuum, not necessarily distinguishable from the spectrum of the noise due to scintillation, extinction variations, sky radiance variations and photon statistics. The proposed correlation technique provides a means whereby systematic errors may be largely eliminated and the full statistical threshold limit achieved. Simultaneous observations of a suspected variable by an array of two telescopes separated by a base line exceeding the noise correlation distance are proposed.

The zero lag covariance estimate $Rxy(0)$ of the irradiance fluctuations $x(t)$, $y(t)$ will have an expectation value $Rxy(0) = \langle x(t) \cdot y(t) \rangle = \sigma_s^2$ equal to the variance intrinsic to the star. The length of the base line required to decorrelate atmospheric noise ranges typically from a few metres for scintillation noise to tens of kilometers for slower extinction and sky radiance changes. The normalised rms error in a zero lag cross covariance estimate is given by $\varepsilon \approx \sigma_n^2/\langle S \rangle^2 (2BT)^{\frac{1}{2}} \sim 10^{-6}(\text{mag})^2$ for $B = 1$ Hz, $T = 10^4$s, where the spectral density of the mean square noise, $\sigma_n^2/B\langle S \rangle^2 \sim 10^{-4}(\text{mag})^2/\text{Hz}$ for a small telescope. The technique is therefore capable in principle of detecting broadband flickering of the order of 10^{-3} mag. We have confirmed the sensitivity of this technique using two telescopes separated by a distance of 5.6 km.

M. Friedjung (ed.), Novae and Related Stars, 53. *All Rights Reserved.*
Copyright © 1977 by D. Reidel Publishing Company, Dordrecht, Holland.

THE BINARY NATURE OF THE NOVA-LIKE VARIABLE CD -42° 14462

A. P. Cowley and D. Crampton

Dominion Astrophysical Observatory

J. E. Hesser

Cerro Tololo Interamerican Observatory

A series of high time resolution spectra reveal the rapidly flickering, nova-like variable CD -42° 14462 to be a binary with P = .206 days and each component near one solar mass. The implied model is very similar to that found for the U Gem stars and old novae: an unseen lower main sequence star is transferring material onto a white dwarf. An accretion disk about the degenerate star is detected in weak H emissions. The phase dependence of both the velocity and intensity of these emissions indicates a hot spot on the following side of the accretion disk. Simultaneous rapid photometry reveals the flickering is not modulated with the orbital period, suggesting the inclination is too low for an eclipse or occultation of the hot spot.

Observations of similar nova-like objects implies that they, like the novae, may all be close binaries. TT Ari has been shown to have an orbital period of .138 days, while BD -7° 3007 (which is currently under investigation by the authors) may have a period near .3 days. In spite of the spectroscopic similarity to old novae and U Gem stars and the overlap in both the mass and period ranges, the nova-like variables are not known to have suffered outbursts. It is proposed that the rate of mass transfer may be lower in the nova-like variables.

RELATION OF THE X-RAY SOURCES SCO X-1 AND CYG X-2 TO OLD NOVAE

A. P. Cowley and D. Crampton

Dominion Astrophysical Observatory

Recent spectroscopic and photometric investigations (Cowley and Crampton 1975; Gottleib et al 1975; Crampton and Cowley 1976) have shown Sco X-1 and Cyg X-2 to be low mass binaries with periods less than a day. Spectroscopically each resembles certain old novae, although in Cyg X-2 the secondary (F) star dominates the spectrum, while for Sco X-1 emission lines from the accretion disk predominate. Both the colors and the rapid flickering in these objects is reminiscent of old novae. Although the masses fall in the usual range for novae and U Gem stars, the periods are about three times the typical values for those stars. In the case of Cyg X-2, one sees spectroscopically that the F star is somewhat evolved, and thus is expanding beyond its Roche lobe as a result of evolution. Although the secondary is not visible in Sco X-1, it is inferred that a similar situation exists. Thus in the low mass X-ray binaries the mass exchange is driven by evolution, resulting in a much higher mass transfer rate than in the case of the novae where the secondary is a lower main sequence star just filling its Roche lobe. It is likely that a further difference is that the X-ray binaries may contain neutron stars rather than white dwarfs as the degenerate member.

Cowley, A. P. and Crampton, D. 1975 Ap J Letters 201, L65
Crampton, D. and Cowley, A. P. 1976 Ap J Letters 207, L171
Gottleib, E. W., Wright, E. L., and Liller, W. 1975 Ap J Letters 195, L33

M. Friedjung (ed.), Novae and Related Stars, 55. All Rights Reserved.
Copyright © 1977 by D. Reidel Publishing Company, Dordrecht, Holland.

MODEL OF U GEMINORUM

Józef Smak

Copernicus Astronomical Center, Warsaw, Poland

Three body approximation is used to describe the rotational motion of the outer parts of the disk and the motion of the stream coming from the secondary component. The family of models, resulting from the numerical calculations, depends on two parameters: the mass-ratio and the dimension of the disk. The radial velocity data, when interpreted within this approximation, lead to: $m_1 = 0.9$ m$\odot$, $m_2 = 0.35$ m$\odot$, $A = 1.0$ x 10^{11} cm, $R_2 = 0.4$ R$\odot$, and the radius-vector of the hot spot $r_s = 0.36A$, with uncertainties up to 10-30 percent.

Rediscussion of the photometric data brings no major changes as compared with earlier results (Acta Astr., 21, 15, 1971) which - whenever comparable - are highly consistent with the new spectroscopic results.

From an assumption that the invisible secondary component is a normal main-sequence star with $L_2 = 10^{32}$ ergs/s, one gets the rate of the mass-transfer dm/dt $> 10^{17}$ g/s and the density of the material in the stream N $>$ 3 x 10^{13} cm^{-3}.

It is shown that a stationary accretion from the disk onto the white dwarf cannot take place between the outbursts and the material coming in from the secondary component must be accumulated in the disk. Its sudden accretion can provide more than enough energy for a typical outburst.

This paper will be published in Acta Astr., 26, No. 4, 1976.

THE CONTINUOUS RADIATION EMITTED BY THE ACCRETION DISKS IN CATACLYSMIC BINARIES. THE DWARF NOVA SS CYG DURING OUTBURST AND THE OLD NOVAE V603 AQL AND RR PIC

Romuald TYLENDA

Institute of Astronomy, Nicolaus Copernicus University, Torun,
Poland

ABSTRACT

A steady-state model of the accretion disk around a white dwarf, including the boundary layer and the hot spot, has been constructed. The calculated distribution of radiation has been compared with the observations of the dwarf nova SS Cyg during outburst and the old novae V603 Aql and RR Pic. The agreement between the observations and the calculations has been obtained for the following parameters of the models :

	SS Cyg	V603 Aql	RR Pic
mass of the primary (10^{33} g)	2.0	1.5	1.5
radius of the primary (10^9 cm)	1.5	0.8	0.8
mass transfer rate (10^{17}g s^{-1})	6.0	8.0	40.0
inclination angle (deg)	36	20	70
distance (pc)	100	376	480
total luminosity of the system (10^{34} ergs s^{-1})	5.4	10.0	50.

M. Friedjung (ed.), Novae and Related Stars, 57. All Rights Reserved.
Copyright © 1977 by D. Reidel Publishing Company, Dordrecht, Holland.

PART II

OBSERVATIONS OF NOVAE AND RELATED OBJECTS
DURING OUTBURST

MODELS OF THE NOVA OUTBURST

M. FRIEDJUNG

Institut d'Astrophysique
98 bis, Bd Arago
75014 Paris, France

Abstract - After a brief review of the observational data for those who are not familiar with the subject, several possible simple models are discussed. After eliminating models in contradiction with the observations, it is concluded that the most probable model is one where most of the mass is ejected near maximum light, but where ejection continues afterwards. This continued ejection would produce in such a case the greater part of the light of the continuous spectrum during certain phases at least, and be responsible for important features of the line spectrum. However many problems of the interpretation still remain.

Introduction - I shall try, in the time available, to introduce what is still a very controversial subject. This talk expresses a personal point of view, and some of you may later wish to challenge what I am going to say. The conception of continued ejection which I support has become more popular in recent years, and I have almost wondered whether this might not be a reason to reject it. In any case, I wish to launch a discussion on phenomena which are still very poorly understood.

Brief Review of Observations

The light curve, which gives the variation of apparent brightness with time, is the most easily obtained property. The light curve in the vast majority of cases includes a very rapid rise in luminosity to about 2 magnitudes below maximum in a time of the order of a day, a slower final rise, and a much slower decline. Light curves differ however very much from each other; the rate of post-maximum decline is far from constant, while secondary maxima and minima of luminosity may occur during the decline. Mc Laughlin (1936) found a common pattern for most novae, and defined a series

of stages, each generally passed through when the brightness was a certain number of magnitudes below that at
maximum. These stages, pre-nova, initial rise, pre-maximum halt, final rise, early decline, transition, final
decline, and post-nova, give however only a very rough
idea of the light variation. The differences in light
curves are very clear if for instance one looks at those
published by Payne-Gaposchkin (1957).

The spectrum of a nova contains strong emission lines, so a light curve of the flux of the continuous
spectrum is more significant than that of integrated
light. Graphs of continuum magnitude against log time
from maximum are particularly interesting. For instance,
such a graph for Nova V603 Aquilae from an old paper of
mine (Friedjung 1966a) shows that late decline occurred
on a line parallel to that of early decline, and the
well known oscillations occurred between the two lines.
The graph for GK Persei shows similar properties.
Straight lines in these graphs indicate that the light
flux varied as a power of the time, and the nature of
the graph for V603 Aquilae makes me very sceptical of
theories which invoke an entirely different process for
early than for late decline. It should be also noted
that graphs of integrated light magnitudes against log
time were first plotted by Vorontsov-Velyaminov (1940).

As far as observations in other parts of the spectrum are concerned, two novae (FH Serpentis and V1500
Cygni) have been observed in the ultraviolet by satellite. The latter which we shall discuss tomorrow seems
to have been rather special, and I shall not discuss it
now. The former was studied by Gallagher and Code (1974)
who showed that during the two months following visual
maximum, the energy distribution shifted to the violet,
while the flux between 1550 and 5480 Å was nearly constant during this period. They concluded that the bolometric magnitude did not change, a conclusion which, as I
shall show in my short contribution, is not necessarily
valid. Infrared excesses for several novae were observed
by Geisel, Kleinmann and Low (1970), explained by the
condensation of dust. In particular FH Serpentis reached
an infrared luminosity peak of $5 \times 10^4 L_0$ about 3 months
after the visual maximum. Radio emission was detected
by Hjellming and Wade (1970), Wade and Hjellming (1971)
and Herrero et al (1971) for FH Serpentis, HR Delphini
and V368 Scuti, and can be interpreted as due to free
free emission from an expanding envelope, which after
being optically thick becomes optically thin.

When one examines the spectra of novae, the lines
during much of the development of a nova have profiles
which are superpositions of several P Cygni type profiles.

Thus a spectrum line can have several blue shifted absorption components, each of which may be at the blue edge of a broadened unshifted emission line. The absorption components belong to different "absorption systems" containing many components of different spectrum lines, with approximately the same blueshift at a given time. This can clearly be interpreted as produced by an envelope in expansion, different parts having different expansion velocities. This very well known fact needs perhaps to be emphasized, as theoretical studies sometimes seem to suppose that all the gas ejected is in one cloud. The absorption systems observed were classified by Mc Laughlin (1943). The "premaximum system" is present before the visual maximum, and disappears usually shortly after, the "principal system" emerges soon after maximum, the faster "diffuse enhanced system" appears usually when the nova has faded about a magnitude and a half, while the yet faster "Orion system" appears still later. The principal and diffuse enhanced systems have as well as absorption components of the Balmer lines, especially components of lines of singly ionized metals. The Orion system has a higher excitation, containing absorption components of oxygen, nitrogen and carbon in various stages of ionization, HeI, but not always the Balmer lines. It should also be noted that when a nova has faded sufficiently, the continuous spectrum and absorption components disappear; the nova then enters the "nebular stage" and has emission lines of the same width as those associated with the principal absorption system. It should also be pointed out that the recurrent novae cannot, it seems, be fitted into this scheme of spectral development.

In fact, the spectral development of a nova is rather less simple than this classical scheme would indicate. It is a means of classifying absorption systems, but more than one system may have the same classification. The blue-shifts of systems having the same classification may be very similar, but this is not always the case. These problems are particularly clear in the case of V 476 Cygni, which had 12 absorption systems, of which 2 were pre-maximum, at least 2 Orion, and many of which were not classified by Payne-Gaposchkin (1957). Also Malakpur (1973) found that the principal system of the very slow nova HR Delphini appeared several months before its main maximum. In view of these considerations, I have wondered for some time, whether we should not be sceptical of the classical scheme.

The temperature and radius of the photosphere of a nova can in principle be determined. By the photosphere, I mean the layer where the gas observed becomes optical-

ly thick in the continuum, i.e. has an optical depth of
about unity. The nature and indeed the existence of such
a layer clearly depends on the model, as we shall see
presently. However it is possible to perform calculations, as if the continuous spectrum come from the surface of a "central star", and its radiation ionized and
excited a surrounding "nebula" responsible for the emission lines. Thus not only can one determine colour temperatures of the continuous spectrum, but also Zanstra
temperatures from the intensities of the Balmer emission
lines, as well as other emission line temperatures.
Knowing the temperature and the brightness of the continuous spectrum, the "radius" of the photosphere can
be determined. One finds, if one performs such calculations, that as the nova fades, the temperatures thus determined rise, while the radii fall, a secondary maximum
in brightness of the continuous spectrum being associated with a fall of temperature, and an increase in the
radius. However it must be emphasized that all the assumptions involved in the determination of these temperatures are certainly not valid, in particular the effect I found during radius oscillations, that the temperature of the photosphere at a given radius tended to
decrease with time (Friedjung,1966b) should be viewed
with a large amount of caution.

These temperature and radius results allow us to obtain more information concerning the absorption systems.
The velocity variations observed were summarized by Mc
Laughlin (1965a). The velocity of the principal absorption system tends to increase at least a small amount
with time, while some diffuse enhanced velocity variations have been noted. It is however theOrion system
which shows the most spectacular variations, showing oscillations in several cases. Now for several well observed novae, these velocities are well related to the calculated photospheric radii. In fact, graphs of velocity^2
against 1/radius are in the cases of a number of well
observed novae, almost linear (Friedjung 1966c). The extrapolation to infinite radius moreover gives in these
cases the velocity of the or one of the diffuse enhanced
systems. This strongly suggests that the same physical
process is behind the Orion and diffuse enhanced systems,
one which, if the ideas I support have some relation to
reality, is associated with continued ejection.

I now come to a point which is crucial for what follows; this is the question of velocity stratification.
Is faster expanding gas nearer the centre, or nearer the
edge of the envelope? It is generally believed to be nearer the centre, as the faster moving gas generally tends
to appear later in the development of a nova, implying

later ejection. Some other reasons for believing this
are given by Mc Laughlin (1965b). Summarizing these reasons, firstly the fastest systems have the highest excitation, which if the photosphere provides radiation for
this excitation, implies line formations close to it.
More decisively diffuse enhanced and Orion absorption
is filled in by overlying principal emission, while
principal absorption absorbs diffuse enhanced emission.
It should be noted however that it is geometrically possible if an envelope possesses a velocity gradient, for
underlying emission to fill overlying absorption when
line blending occurs. The stratification, generally assumed, if true, enables one to eliminate many possible
models, as we shall see. Indeed when I am depressed, I
sometimes wonder whether it helps to eliminate all possible models. In any case, if the stratification could
be reversed, and an "inside out" model adopted, the situation would be much simpler.

Models -
     Having briefly considered some of the more signifi-
cant basic observational facts, we can now consider possible models. The models to be considered are of stars
which eject gas during a limited time, so as to produce
a tempory increase in brightness. Five rather simple types of model exist, which I described two years ago,
(Friedjung 1974). These models are as follows :

Instantaneous ejection models :
     In such models all or nearly all the gas is ejected
in a time short compared with that in which a nova undergoes its characteristic development after maximum light.
The changes observed during this development are those
of previously ejected gas. Two simple forms of instantaneous ejection can be considered :
     Instantaneous ejection type I -
The gas is ejected in a thin shell across which the outward expansion velocity does not vary much. In the case
of a nova, this would have to be the gas associated with
the principal absorption system, which contains a large
part of the mass of the envelope, as we shall see. A model of this type was extensively studied by Pottasch,
(1959 a,b,c,d).
     Instantaneous ejection type II -
The ejection velocity of different parts of the envelope
is not the same. The result is that instead of remaining
in a thin shell, the ejected gas fills a larger volume
with the fastest moving gas near the outer edge, and the
slowest near the centre, wide absorption lines would be
produced in this case. A model of this type is described
by Sobolev (1958), in which the velocity of each mass of

gas does not change after ejection, so the distance tra-
velled from the centre of the envelope is proportional
to the velocity. This type of model is used to interpret
supernova spectra. An instantaneous ejection type II model was also suggested by Nariai (1974) for ordinary
novae, taking into account gravitational deceleration,
which is especially important for the slower moving gas
near the centre.

Continued ejection models

In models of this type ejection continues after the
visual maximum, but generally tends to decrease with time. In the simplest situation which can be envisaged,
the envelope then tends to have two density maxima. One
is associated with gas ejected in the early stages when
the ejection rate is high, and occurs near the outer
edge. The other maximum, which exists as long as ejection continues, and even a short time after, is due to
newly ejected gas, which has not expanded much. Two simple forms of continued ejection occur.

Continued ejection A

The central density maximum contributes to most of the
observed light, especially of the continuous spectrum.
The photosphere is formed in layers of expanding gas,
and its properties are those of spherically symmetrical
extended atmospheres. This is the type of model I have
genrally supported in my papers.

Continued ejection B

The outer density maximum contributes to most of the
light, even in the case of the continuous spectrum. The
properties of this kind of model are very similar to
those of instantaneous ejection type I.

Central Star Dominant Models

In models of this kind, most of the light of the
continuous spectrum comes after visual maximum from a
region not in expansion. This region could be the surface of one or both the central stars, or on expanded almost stationary envelope. A model where the cooler secondary star is heated by X-rays from the primary, as
has been suggested by Endal et al (1976) and by Wu et
al (1976) for the transient X-Ray source AO 620-00 =
Nova Monocerotis 1975, is of this type. In the case of
a classical nova however, the photosphere during the
early decline in light is much larger than the undistur-
bed radii of the central stars, so expansion of at least
parts of one of the stars would be required. A model involving an expanded central star has been supported by
Mustel (1957). It should also be mentioned that theoretical studies (see for instance Starrfield et al (1974)
suggest expansion of the hot stellar component,of the
binary system, to wich a nova belongs.

Tests of Models -

   Most of these models, at least in their simple form,
are contradicted by the observations. Firstly the visual
rate of fading of V 603 Aquilae at times when the hydrogen was probably completely ionized is too rapid for continued ejection or instantaneous ejection type I, at
least in their elementary formulation.

   The study of the spectra gives stronger arguments.
An instantaneous ejection type II model requires lower
velocities near the centre of the envelope, in contradiction with what has been said about velocity stratification. It showed be noted that with such a model apparent continuing activity could be explained if the centre of the envelope was denser than the outer parts, and
optically thin later. However in this case, the emission
line widths would decrease with time, as slower moving
gas became visible and dominated the spectrum. This is
not observed for classical novae while such an effect
though observed in the cases of the recurrent novae RS
Ophiuchi and T Coronae Borealis, can be explained then
by the deceleration of an expanding shell due to a stationary circumstellar cloud (Pottasch 1967, and Gorbatskii 1972).

   There are reasons to believe that most of the continuous spectrum comes from regions near the centre of the
envelope, contradicting instantaneous ejection type I
and continued ejection B. The narrowness of the principal system absorption components of V 603 Aquilae formed, taking into account what has been said about stratification far from the centre, when studied closely,
is difficult to understand unless the continuous spectrum was formed much closer to the latter. Moreover, a
large part of the continuous spectrum of this, and indeed of other novae also, was formed at least at certain
times, even nearer the centre than the high velocity
Orion system. This conclusion follows from the strength
of line absorption due to NIII. A similar result for the
region of emission of the continuous spectrum was obtained from an argument based on the intensity of hydrogen
absorption before the December 1967 maximum of HR Delphini in a study in which I participated (Friedjung and
Malakpur 1973), but the argument is unfortunately not
watertight.

   There are also difficulties with central star dominant
models. If they described the situation which occurs,
one might expect to see traces of unshifted spectrum lines, formed in or near a non expanding photosphere. Now
if one excludes the interstellar lines, a faint undispla-
ced absorption spectrum may have briefly existed in the
cases of DN Geminorum, GK Persei, and perhaps also V603

Aquilae, according to Mc Laughlin (1943), who considered
it circumstellar. One would, I think, expect a more systematic appearance of such lines, if they were due to a
stationary photosphere, while the study of FH Serpentis,
about which I shall talk in my short contribution, indicates that such absorption lines were missing. Narrow
emission lines, which also must come from a non-expan-
ding region, were observed for the recurrent nova RS
Ophiuchi, and convincingly explained by Pottasch (1967)
as circumstellar. Relatively narrow parts of emission
line profiles of HR Delphini were interpreted as due to
the heated atmosphere of a companion star in a study I
partly performed (Friedjung and Malakpur 1971), but I
now think this interpretation wrong, because of the lack
of infra-red radiation which would have been produced by
the companion in 1970. Hutchings (1969) considered that
all the line emission of this star came from the same
expanding region. What I have stated does not contradict
a possible expansion of the star responsible for the explosion; I only do not think that we can see it during
the active phases of a nova.

   As you will have noticed, I have criticized all models
except continued ejection A. In fact, I have not been
completely honest. Though this model is in my opinion
closest to the truth, it also has a number of difficulties. According to it, the rate of ejection at any time
is closely related to the brightness of the continuous
spectrum, so the variation of the latter directly gives
information concerning the variation of the former. It
must also be stated that, according to plausible forms
of the model, continued ejection is associated with the
Orion and diffuse enhanced absorption systems. This is
because these systems seem to be nearer the centre of
the envelope and hence the photosphere after visual maximum. The Orion system, in particular, shows large velocity variations, which one may suppose associated with
changing ejection conditions, while the principal system
generally has a much more regular velocity behaviour.
This means that one may expect collisions between continuously ejected gas of the Orion absorption system and
slower moving gas further out, particularly of the principal absorption system. These considerations then lead
to conditions which are not easy to reconcile, and which
are :

   1) Calculations I performed (Friedjung 1966b) for several novae using photosphere radii calculated from temperatures derived by Zanstra's method and supposing the
photosphere radiated like a grey extended atmosphere according to the simplified theory of Kosirev (1933), suggested a slow decrease of ejection rate with time.

2) The increase in velocity of the principal system, which should occur as a result of the collision of continuously ejected gas with that responsible for the principal systems, is often not large. This suggests that most of the mass of the envelope is ejected near the visual maximum, when this system becomes visible.

3) The conclusion that most mass is ejected near maximum light is supported if one considers the emission line profiles. During the nebular stage, the width corresponds to the Doppler broadening of gas with the velocity of the principal system, though weak Orion wings may exist. In this stage of course, all the high velocity gas may have been destroyed by collisions with slower moving gas. Even during earlier stages, however, Orion emission may be weak.

One can attempt to escape from these apparent contradictions using rather artificial models such as models involving continued ejection at the principal absorption system velocity, or by invoking departures from spherical symmetry, as such departures are known to exist. However, it appears to me now that the most rational explanation is that for many, if not most novae, the situation is in fact a cross of continued ejection A and instantaneous ejection type I. By this I mean that as a nova declines in light, most radiation of the continuous spectrum comes from a photosphere produced by continuously ejected gas, while most of the mass is contained in the gas of the principal absorption system, ejected near maximum light.

In order to see how this can be, let us reconsider the apparently contradictory conditions, which I have just described. The rate of decrease of the ejection rate found from the photospheric radii is clearly unreliable. Calculations should be based on proper non grey non LTE. spherical model atmospheres, with a density which varies to a first approximation as the inverse square of the distance from the centre. Such models including non LTE effects have not yet been calculated as far as I am aware. The weakness or near absence of Orion emission which can occur, similarly may be seen to be not significant, when the correct calculations have been carried out. There are also two effects which in particular may have led to an overestimate of the ratio of continuously ejected mass to that ejected near maximum. These are :
a) near visual maximum the temperature is rather low, and most hydrogen in the photosphere may be neutral. This means the continuous absorption can be much lower than that I assumed, and hence the ejected mass for a given

photospheric radius larger. However there are limits on
the mass, one can suppose to be ejected near maximum
light, as the mass of the envelope can be estimated in
the nebular stage, and according to Pottasch (1959c) is
of the order of 10^{-5} to 10^{-4} $M_\odot$.
b) If rapid variations in the rate of ejection occur
during the period of continued ejection, a lower mass
of the region near the photosphere may correspond to the
same continuum optical thickness. If the mass absorption
coefficient is proportional to the density, a smaller mass
is required if it is concentrated in thin slabs, for a
certain optical thickness to be attained. In this case,
if the thickness in the radial direction is T, the mass
required for a constant optical thickness is proportional to $\sqrt{T}$. However electron scattering may dominate,
and in this case the condition for a photon to have a
certain probability of escaping without suffering pure
absorption is, if the latter is still proportional to
density, proportional to $\sqrt[3]{T}$.

It should also be noted that the results of Pottasch
(1959c) support the kind of model now suggested. He calculated the mass of the envelope by two methods assuming instantanious ejection type I, and obtained consistent results in spite of the very approximate nature of
the calculations. This can be explained if most hydrogen
line emission comes from the gas ejected near visual
maximum.

There still however remains another question concerning the model, I suggest, as well as many other possible models. Collisions can be expected to occur with
relative velocities of the order of 1000 Km/sec, giving
rise to X ray emission. A collision at this velocity
would assuming an efficiency factor of 0.1 characteristic of a strong shock, give rise to a temperature of
1×10^{7} ° after thermalization of the flow. Optically
thin free free emission at this temperature would decrease sharply below 15 Å. The total luminosity due to
collisions according to the model suggested here might
be of the order of 10^{36} erg/sec during the continued
ejection phase. The flux for energies above 2 Kev would
be an order of magnitude lower, while the optical depth
for absorption could be as much as 10^{5} at 1 Kev around
visual maximum and still 10^{2} a month later. Thus the fai-
liure to detect ordinary novae as transient X ray sources can perhaps be explained. According to Silk (1973)
there are not enough transient X ray sources for them
to be identified with novae, while Hoffman et al (1976)
showed that in the case of V 1500 Cygni, the ratio of
1.5 - 15 Kev luminosity to optical luminosity was not
more than 10^{-4}. However V 1500 Cygni, which we shall

discuss tomorrow was not normal, having probably little or no continued ejection. In the case of ordinary novae, the question of the predicted X ray emission can only be settled with more detailed calculations based on detailed models.

Accepting the model, I am now suggesting, one can ask whether there is a simple interpretation of the different absorption systems. The premaximum system most probably generally consists of gas ejected before the visual maximum, and swept up by later ejected gas. The gas of the principal system which becomes visible at maximum can be identified with the outer peak of continued ejection models, ejected when the ejection rate is high. The other systems may not always be produced by the same phenomenon. However in some cases at least (eg. V603 Aquilae), it is tempting to associate the Orion system with gas near the photosphere where the hydrogen is nearly completely ionized by Lyman continuum radiation, but where the relatively high density produces absorption lines. The diffuse enhanced system could them be associated with gas just outside the HII region during phases when the latter was small and confined to regions near the centre of the envelope, where the density would vary as the inverse square of the distance from the centre. The region just outside the HII region would then have a density maximum of neutral hydrogen and singly ionized metals, quite distinct from the outer maximum. The diffuse enhanced system of V603 Aquilae which only lasted 5 days, could have been associated with such an effect. It may be noted that a different explanation of absorption systems, based on deviations from spherical symmetry was given by Hutchings (1972).

I suggested some years ago, (Friedjung 1971), that ionization from level 2 of hydrogen, was important in the regions of formation of the principal and diffuse enhanced absorption systems. Though the reasoning of this paper needs revison, there is no doubt that ionization from level 2 occurs, as one sees Balmer absorption lines, and there is a finite (of order 10^{-2}) optical thickness in the Balmer continuum. I probably underestimated the proportion of neutral hydrogen in the region of the principal system in that paper, but even if most of the hydrogen of this system is neutral, I do not believe this to be always the case for the diffuse enhanced system. In any case the ionization conditions can only be definitely determined from combined detailed observational and theoretical studies.

Before concluding this review, I should mention that the radiation flux is probably above the Eddington limit, during at least some phases after visual maximum, as I

shall show in my short contribution for FH Serpentis.
This is a necessary though not a sufficient condition,
if one wishes to identify continued ejection with a radiation pressure driven wind as suggested by me (Friedjung 1966c), as well as more recently by Bath and Shaviv
(1976).

In conclusion, I wish to say that the model I suggest
appears to me the most probable one for many novae. However there are still numerous problems which can only
be resolved if detailed observational and theoretical
studies are undertaken. It is necessary to study novae,
which often are very different from each other as indi-
viduals and not over generalize. These problems have
worried me for a number of years, but perhaps we are
now somewhat nearer the solution.

REFERENCES-

Bath,G.T., Shaviv,G. 1976, Mon. Not. R. Astr. Soc. 175,
          305.
Endal,A.S., Devinney,E.J., Sofia,S. 1976, Astrophys.
          Letters 17, 131
Friedjung,M. 1966 a, Mon. Not. R. Astr. Soc. 131, 447.
Friedjung,M. 1966 b, Mon. Not. R. Astr. Soc. 132, 143.
Friedjung,M. 1966 c, Mon. Not. R. Astr. Soc. 132, 317.
Friedjung,M. 1971, Astr. Astrophys. 14, 440.
Friedjung,M. 1974, "in variable Stars and Stellar Evo-
          lution", Ed. Sherwood and Plaut, IAU Sump.
          67, p. 335.
Friedjung,M., Malakpur,I. 1971, Astrophys. Letters,7,171.
Friedjung,M., Malakpur,I. 1973, Astrophys. Space Sci.
          25, 433
Gallagher,J.S., Code,A.D. 1974, Astrophys. J. 189, 303.
Geisel,S.L., Kleinmann,D.E., Low,F.J. 1970, Astrophys.
          J. 161, L 101
Gorbatskii,V.G. 1972, Astr. Zu. 49 , 42.
Hjellming,R.M., Wade,C.M. 1970, Astrophys. J. 162, L1
Hoffman,J.A., Lewin,W.H.G., Brecher,K., Buff,J., Clark,
          G.W., Joss,P.C., Matilsky,T. 1976, Nature 261,
          208.
Herrero,V., Hjellming,R.M., Wade,C.M., 1971, Astrophys.
          J. 166, L 19.
Hutchings,J.B. 1969, in "Mass Loss from Stars", Ed. Hack
          (Reidel) dordrecht - Holland), p. 290.
Hutchings,J.B. 1972, Mon. Not. R. Astr. Soc. 158, 177.
Kosirev,N.A. 1933, Mon. Not. R. Astr. Soc. 94, 430.
Malakpur,I. 1973, Astrophys. Space Sci. 24, 577.
Mc Laughlin,D.B., 1936, Astr. J. 45, 145.
Mc Laughlin,D.B. 1943, Publ. Obs. Univ. Michigan 8, 149.
Mc Laughlin,D.B. 1965 a, In"Novae, Novoides et Superno-

vae" CNRS Paris, p. 3
Mc Laughlin,D.B. 1965 b   , in"Novae, Novoides et Su-
            pernovae" CNRS Paris, p. 123.
Mustel,E.R. 1957 in "Non Stable Stars" Ed. Herbig, IAU
            Symp. 3, p. 57.
Nariai,K. 1974, Astr. Astrophys. 36, 231.
Payne-Gaposchkin,C.H. 1957 "The Galactic Novae" (North
            Holland Publishing Company, Amsterdam).
Pottasch,S.R. 1959 a, Ann. Astrophys. 22, 297.
Pottasch,S.R. 1959 b, Ann. Astrophys. 22, 310.
Pottasch,S.R. 1959 c, Ann. Astrophys. 22, 394.
Pottasch,S.R. 1959 d, Ann. Astrophys. 22, 412.
Pottasch,S.R. 1967, Bull. Astr. Inst. Netherl. 19, 227.
Silk,J. 1973, Astrophys. J. 181, 747.
Sobolev,V.V. 1958, in "Theoretical Astrophysics" Ed.
            Ambartsumian (Pergamon Press, London),p.455.
Starrfield,S., Sparks,W.M., Truran,J.W. 1974, Astrophys.
            J. Suppl. Ser. 28, 247.
Vorontsov-Velyaminov,B.A. 1940, Bull. Inst. Sternberg,
            Moscow 1.
Wade,C.M., Hjellming,R.M. 1971, Astrophys. J. 163, L 65.
Wu,C.C., Aalders,J.W.G., van Duinen,R.J., Kester,D.,
            Wesselius,P.R. 1976, Astron. and Astrophys.,
            in press.

THE EXPANSION OF NOVAE BEFORE LIGHT MAXIMUM AND THE FORMATION OF THE PRINCIPAL ENVELOPE

E.R.MUSTEL

Astronomical Council, USSR Academy of Sciences,
Moscow, U.S.S.R.

There are many reasons to state that the principal envelope of Novae usually is formed just immediately after the moment of light maximum, t_{max}. Therefore in order to understand the mechanism which is responsible for the origin of the principal envelope we should analyse all the events which are observed from the moment of explosion of a Nova up to the moment of appearance of the principal spectrum of the star. At first we should mention that already for three Novae we have information on their spectra, obtained before the explosion by means of objective prisms, see discussion of this in paper of Arhipova and Mustel (1975).

Unfortunately till now we had no chance to record any nova during the first stage of its outburst. At the same time for a few Novae we have more or less complete photometric and spectroscopic information before light maximum, at light maximum and after it. Let us consider at first the rapid Novae.

The spectra and U,B,V-photometry before light maximum show that the temperature of the star is decreasing during this period. Furthermore available spectrophotometric measurements indicate that the energy distribution in the spectra of these objects is close to the black body radiation. Thus we may compute the change of the "photospheric" radius R_p of the expanding star before t_{max}, using, for example, the following very well known expression:

$$M_v = \frac{29500}{T} - 5\log R_p - 0.08 \tag{1}$$

M. Friedjung (ed.), Novae and Related Stars, 75-85. All Rights Reserved.
Copyright © 1977 by D. Reidel Publishing Company, Dordrecht, Holland.

where M_v is the absolute magnitude of the Nova and T is the temperature of the star, estimated according to its spectral class or on the base of spect - rophotometric measurements. The validity of appli - cation of (1) to Novae is discussed in detail in the papers of Mustel (1945).

On the other hand we may compute the growth of the radius of those expanding outer layers of the Nova which produce its absorption spectrum by integrating the velocity V_D deduced from Doppler dis - placement of absorption lines of the spectrum of Nova:

$$R_D = R_o + \int_{t_o}^{t} V_D \, dt$$

(2)

where R_o and t_o determine the initial conditions of the integration.

The principal observational facts related to this subject and the results of computations carried out on the base of (1) and (2) may be summarized as follows:

a) The temperature of the Nova before t_{max} is continuously decreasing and stops to decrease at light maximum. After this the temperature of the star begins to increase with some fluctuations. And it is interesting to note that during the subsequent secondary fluctuations of brightness the temperature of the Nova is low at the maxima of brightness and increases at the minima of brightness.

b) According to available, though limited, data the velocity V_D is decreasing towards t_{max} , see paper of Beer (1974) for DQ Her.

c) According to computations of Beer (1937) carried out for DQ Her the behaviour of the values of R_P and R_D for the first observational period was quite different. Namely the magnitude of V_P during several days after the discovery of the Nova (Dec. 13, 1934) was considerably smaller than V_D . This led to the formation of a very extended "reversing layer" around the "photosphere" and generally the character of variations of R_P and R_D was quite different.

d) It was concluded in the papers of Mustel (1945) that immediately before light maximum every Nova possesses, similar to DQ Her, a very extended outer envelope responsible for the absorption spectrum of the star. The linear thickness of this envelope may exceed several times the "photospheric" radius R_P . Besides it was also found (Mustel (1946)) that during the last 24 hours before t_{max} the average velocity V_P attains its maximum magnitude and becomes approximately equal to the velocity V_D given by the pre-maximum absorption spectrum of the star for the same period. Table 1 illustrates these results.

THE STAR	M_v	$R_P/R_\odot$	V_P	V_D
V 603 1918	-9.2	331	1700	1300
V 476 1920	-8.9	229	733	400
RRPic 1925	-7.3	224	250	285
DQ Her 1934	-5.8	96	185	176
CPLac 1936	-9.2	276	1100	1300
V 1500 Cyg 1975	-10.9	465	1900	1800

Table 1

The fact that the velocity V_P of the "photospheric" layer of Novae is the greatest just before light maximum may play an important role for the problem of formation of the principal envelope. It is very interesting that in the case of the slow Novae the photospheric velocity V_P shows also the greatest magnitude during the last day before light maximum (See further). It is worth mentioning that during this period (for the fast and for the slow Novae) the expanding photosphere is steadily cooling and there are no peculiar variations in the speed of this cooling which might indicate the presence of some shock waves.

e) Finally we may mention that some Novae after their discovery showed (especially DQ Her) relatively strong emission in their spectra. This was due to relatively high temperature of the star at this time and to the presence of an extended envelope. At light maximum (the lowest temperature of the star) the emissions are usually very weak.

Now let us consider the slow Novae. These stars are characterized by relatively slow changes of brightness. Fig.1 shows the changes of the following parameters of a slow Nova HR Del, see Antipova (1977): a) visual brightness m_v; b) temperature T estimated on the base of magnitudes of B, V; c) photospheric radius R_P, estimated by using the expression (1); radial velocities, calculated on the

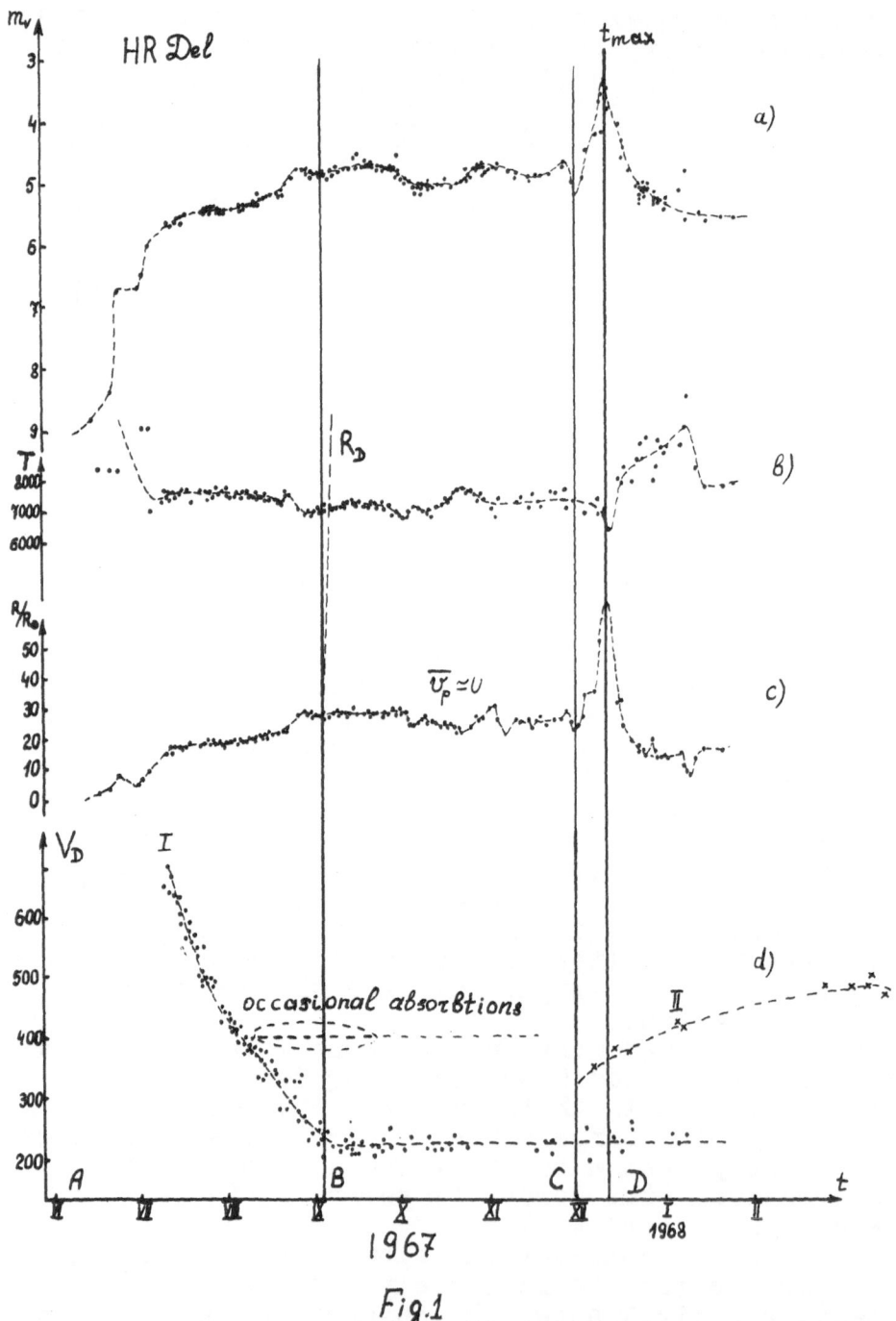

Fig. 1

base of many spectrograms. Fig. 1 confirms the conclusions of the previous studies (see for example Friedjung and Malakpur (1973)) that before light maximum
there was a continuous ejection of gases from HR Del.

In fact we see from Fig.1c, that during the period
$B-C$ the photospheric velocity V_p was practically
zero, on the average, whereas the Doppler displacement during this period was quite large and corres-
ponded approximately to 220 km/sec. It was found by
Antipova that many "sporadic" absorptions in the
spectrum of HR Del were due to erratic ejections of
relatively dense gaseous clouds from the Nova, which
persisted for many days.

Again as in the case of fast Novae we see in
Fig.1c that immediately before light maximum t_{max}
there was a noticeable increase of the "photospheric" velocity of HR Del together with a drop of
temperature. A similar noticeable increase of "photospheric" velocity just before light maximum was
observed for another slow Nova RR Pic, see Mustel
(1946). And it seems that in general this increase
of V_p just before the moment t_{max} is a very common
property of all Novae, fast and slow. Correspondingly there are all reasons to consider that the formation of the principal envelope in the case of HR Del
took place at the moment t_{max} (Fig.1), but not during the first moments of the evolution of the exploded star.

In connection with the same problem it is important to note, that there is a distinctive property
of slow Novae - a continuous ejection of gases from
the star before t_{max} even during the periods of approximate constancy of the photospheric radius. On
the contrary in the case of fast Novae the magnitude
of V_p before light maximum is always rather large.

The next very important fact which may throw
light upon the mechanism of formation of the principal envelope are the following phenomena which are
observed most frequently immediately after light maximum (McLaughlin (1943, 1960)):
a) The replacement of the pre-maximum absorption
spectrum of the Nova by a new absorption principal
spectrum which is accompanied (sometimes with a
small delay, DQ Her) by an emission. From its beginning the absorption system of the principal spectrum
appears to be detached from the absorption system of
the pre-maximum one, in more detail see paper of
Mustel (1957). It is very important to note that the
expansion velocity deduced from the displacement of
the principal absorption spectrum coincides with the
velocity which corresponds to the half-widths of
emission bands of the spectrum of a Nova during its
nebular stage. Thus the transformation of the pre-

maximum spectrum of a Nova into the principal spectrum corresponds to the "birth" of the principal en -
velope and this "birth" takes place immediately af -
ter light maximum. In some rare cases, for example
in the case of RH Del, Fig.1, the first "hints" of
the presence of the principal spectrum are observed
shortly before t_{max}. The replacement of the pre-maximum spectrum by the principal spectrum we shall
call the post maximum transformations, or shortly
PMT.
b) Very soon after the moment t_{max} (sometimes even
a few hours after t_{max}) there appear two quite new
sbsorption systems in the spectrum of the Nova, the
diffuse-enhanced and the orion spectra with displa-
cements higher than the displacement shown by the
principal spectrum. Both these absorption systems
are also accompanied by emissions. The level of excitation and ionization of atoms which produce these
two new systems is higher (especially in the case of
the Orion spectrum) than the level of excitation and
ionization of atoms which produce the principal
spectrum. According to McLaughlin (1943, 1960) these
two absorption and emission spectra are due to a
continuous ejection of gases from the Nova after the
moment t_{max}. The same problem of continuous ejection
of gases from Novae after t_{max} which produces the
diffuse-enhanced and orion spectra was discussed
from the physical point of view in papers of Mustel
(1947). It was concluded in this paper that the clouds which produce the diffuse-enhanced spectrum are
ejected from the superficial layers of the contracting Nova, placed just above the "photosphere", whereas the condensations which produce the orion spectrum are ejected from the relatively unstable regions of the Nova localized somewhere below the "photosphere" of the star. This conception is in agreement with the fact (McLaughlin (1943)), that the
orion spectrum of certain Novae do not contain hydrogen lines. Thus it may be suggested that the gases
which produce the orion spectrum are ejected from
the layers which are close to the outer surface of
the white dwarf.
    In connection with the problem of the continuo-
us ejection of gases from the internal layers of Novae the present author paid attention to very inte -
resting changes in the form of emission profiles in
the spectra of V 603 Aql after light maximum. During
the s e c o n d a r y minima of brightness these
profiles had two "horns" with a deep minimum between

them, whereas during the secondary maxima this minimum disappeared, see Sayer (1935). It was suggested by Mustel (1949,b) that these phenomena were due to a changing intensity of the continuous ejection of gases from the "subphotospheric" layers of the Nova. During the secondary minima (increased temperature of the star) the continuous gas ejection was the least and the bright "horns" were due to fluorescence phenomena in the dense "polar caps" of the star (see further). At the same time during the secondary maxima the continuous ejection of gases was the strongest and the line emission from these gases filled in the central parts of the emission hydrogen profiles of the spectrum of V 603 Aql. Correspondingly it was suggested (in the same paper) that after light maximum there was no usual hydrostatic equilibrium in the Nova V 603 Aql and therefore there are reasons to state that in general the configuration of every Nova after light maximum is determined mainly by such processes which may be considered as a continuously lasting explosion. And finally it was concluded that the outer "photospheric" layers of the star are maintained by the continuous "attacks" of gases ejected from the more internal parts of the Nova(situated somewhere above the outer layers of the white dwarf).

Thus from all the above said we have all reasons to state that the moment of light maximum of a Nova is not a simple "optical" moment, when the star attains its maximum brightness. On the contrary this moment has a g r e a t p h y s i c a l m e a n i n g and sharply defines two totally different stages in the evolution of the explosion of the Nova.

Since the first significant event in the spectra of Novae after t_{max} are the phenomena of PMT we shall consider the origin of these phenomena. A summary of different attempts to explain PMT may be found in papers of Mustel (1949a, 1957, 1962).

One of the models explaining the phenomena of PMT, which was suggested by Mustel (1957) is shown in Fig.2. According to this model the "photosphere" of the Nova having attained its greatest size begins (immediately after light maximum) to contract. At the same time the uppermost layers of the Nova leave the star. An acceleration of these layers by a force F after light maximum produces the phenomena PMT, see in more detail the paper of Mustel (1962). It is suggested in this paper that the force F which produces the acceleration is the pressure of the ener-

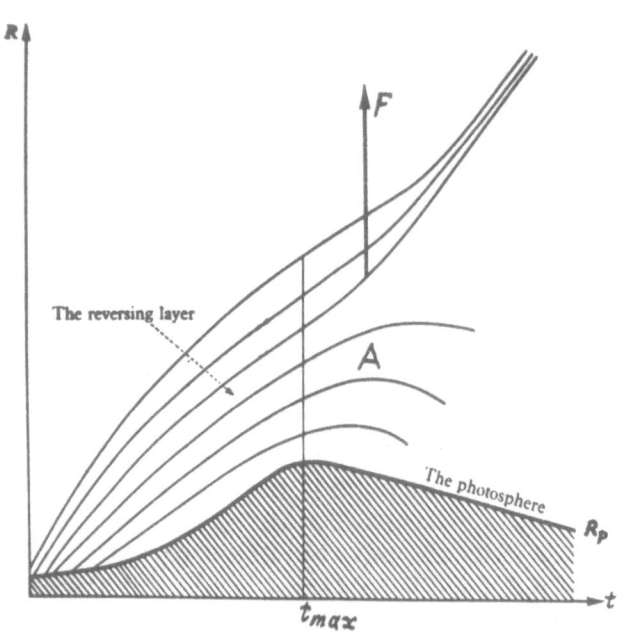

Fig. 2

getic particles which may originate in the "cavity" A between the contracting photosphere and the receding parts of the outer envelope. This of course suggests the presence of magnetic fields in the envelopes of Novae, see further.

A very high level of instability in the above cavity A follows from the absence of absorption lines which are expected to originate in the spectrum of an envelope adjacent to the contracting "photosphere". This instability may be later a source of the diffuse-enhanced spectrum.

As to the origin of the condensations which produce the orion spectrum of the Nova we may suppose that the forces which eject these condensations outwards begin to act from the first moments of the explosion of the star but that they are able to overtake the "photosphere" only after the moment when it stops to expand. Then the condensations of the continuous ejection acquire the possibility to leave the "photosphere".

Of course we might consider an opposite situation according to which the primary factor are the clouds of the continuous ejection and that they accelerate the photospheric layer of the Nova especially during the last days of its expansion. But in this case there are many questions. For example we know that in certain cases the orion spectrum (as well as the diffuse-enhanced spectrum) appears only a few days after light maximum. At the same time the phenomena of PMT begin almost always immediately after light maximum. Then, in this case we cannot understand the fact why the velocity V_p is the greatest just before light maximum and so on.

According to the model shown in Fig.2 the "photosphere" of the Nova does not leave the star. And since the velocity V_p and the radius R_p before light maximum are relatively large it means that in the envelope of the Nova there must exist certain forces which exceed considerably the gravitational forces. In this connection it was suggested (Mustel (1957, 1958)) that Novae possess strong magnetic fields of a dipolar nature, though a certain entanglement of the magnetic field strength lines may exist. Taking into account a model of Novae proposed by Crawford and Kraft (1956) as well as the well known fact that many white dwarfs possess strong magnetic fields the present author suggested a model which explains the origin of strong magnetic fields in Novae, see the review of Arhipova and Mustel (1975). Another confirmation of the hypothesis of strong magnetic fields is the discovery in DQ Her of a strong magnetic field, see Swedlund et al. (1974), Kemp et al. (1974) as well as Lamb (1974).

The presence of strong magnetic fields of a dipolar type in Novae may explain such peculiar features as the "polar caps" in the nebulae around Novae, see Mustel (1957, 1958, 1970), as well as the papers of Mustel and Boyarchuk (1970) and Arhipova, Mustel (1975). It seems that these "polar caps" are very small gaseous formations. For example even 22 years after the explosion of V 603 Aql one of the polar caps of the nebula around the star was very small, see Fig.4 in the paper of Mustel and Boyarchuk (1970). The other, receding polar condensation was noticeably weaker, but the spectra of the nebula showed clearly its presence.

Now about the problem of deceleration (retardation) of gases inside the envelopes of Novae by the magnetic fields of these objects. It is quite clear that in the hydrogen-rich regions close to the outer boundary of the white dwarf (very large magnitude of g) the principal forces must be the gravitational forces and we should speak only about more or less outer layers of the expanding envelope of the Nova.

Let us consider the outer envelope of DQ Her, to be more exact a region A, situated just above the photosphere of the expanding star before light maximum. For the radius R_p which corresponds to this region A we may accept according to Mustel (1946) the value of about 100 $R_\odot$. Now we accept according to Lamb (1974) that the magnetic field strength H_o of

the white dwarf inside DQ Her is of the order of 10^8
gauss. Since the radius of DQ Her in its "normal"
state is of the order of $0.1R_\odot$, therefore the magnetic field strength in the region A will be of the
order of 100 gauss (accepting $H \sim R^{-2}$). Now on the base of available estimates we may accept that in this
region $n_e \simeq n_p \simeq 10^{13} cm^{-3}$ and therefore $\rho \simeq 10^{-11} g$. At
last the mean velocity of the "photospheric" gases
in the envelope of DQ Her was of the order of 175
km/sec. Using all these data we find that inside the
region A the magnitude of $H^2/8\pi$ is only a little less
than $\frac{1}{2}\rho V^2$. The equality between these quantities is
realized for $H \simeq 4 \times 10^8$ gauss. However there are reasons to consider that the average density of the ga-
ses in the envelopes of Novae may be considerably
lower (due to a "patchy" structure of the envelopes)
than the "spectroscopic" value $n_e \simeq n_p \simeq 10^{13} cm^{-3}$ which
we accepted, see paper of Mustel and Antipova (1971).
In this case $H^2/8\pi > \frac{1}{2}\rho V^2$.

We outlined briefly some aspects of the problem
of the origin of the principal envelope. But there
is also a special question of the fine structure of
the envelopes - nebulae around Novae. I spoke already about such peculiar details as the "polar caps"
in the envelopes. There is another important property of the envelopes - the relatively bright belts.
The axis of the symmetry of these belts coincides
approximately with the axis passing through the "polar caps" of the envelope. It seems that this common
axis coincides also with the axis of the rapid rotation of the Nova, see the review of Arhipova and
Mustel (1975). It may be suggested that the above
mentioned belts are due to the influence of different
factors, among which the most important are: the very rapid rotation of the Nova (for DQ Her, $P = 142 sec$)
and the deflecting action of the magnetic field of
the Nova on the outflowing plasma.

References

Antipova L.I., 1977, Astron. Zourn. 54, № 1.
Arhipova V.P, and Mustel E.R., 1975, "Variable Stars
        and Stellar Evolution", IAU Symp. № 67, 305.
Beer A., 1937, Monthly Notices R.A.S., 97, 231.
Beer A., 1974, Vistas in Astronomy, 16, 179
Crawford J.A. and Kraft R.P., 1956, Astrophys. J.,
        123, 44
Friedjung M. and Malakpur I., 1973, Astrophysics and
        Space Science, 25, 433

REFERENCE -

KEMP,C.J., SWEDLUNG,J.B. and WOLSTENCROFT,R.D. 1974,
        Astrophys. J.,193, L 15
LAMB,D.Q. 1974, Astrophys. J. 192, L 129
MUSTEL,E.R. 1945, Astron. Journ. of Sov. Union, 22, 65
        and 185
MUSTEL,E.R. 1946, Astron. Journ. of the Sov. Union, 23,
        289
MUSTEL,E.R. 1947, Astron. Journ. of the Sov. Union, 24,
        97 and 155
MUSTEL,E.R. 1949 a, Publ. Crimean Astrophys. Obs. 4,152
MUSTEL,E.R. 1957, "Non-Stable Stars", IAU Symp. n°3, 57
MUSTEL,E.R. 1958, "Electromagnetic Phenomena in Cosmical
        Physics", IAU Symp. n°6, 193
MUSTEL,E.R. 1962, Astron. Zhourn., 39, 185
MUSTEL,E.R. 1970, Astrophys. Letters, 6, 207
MUSTEL,E.R. and BOYARCHUK,A.A., 1970, Astrophysics and
        Space Sc. 6, 183
MUSTEL,E.R. and ANTIPOVA,L.I. 1971, Scientific Informa-
        tion of the Astronomical Council, 19, 32
SAYER,A.R. 1935, Harward College Observ. Circular, n°406
SWEDLUND,J.B., KEMP,J.C. and WOLSTENCROFT,R.D. 1974,
        Astrophys. J. 193, L 11

THE VARIABLE RADIAL VELOCITY OF HE I 5876 IN THE SPECTRUM OF NOVA DELPHINI 1967 /HR DEL/.

Eugeniusz Szumiejko

Wrocław University Observatory /Poland/

The study has been based on the spectroscopic material obtained between July 20, 1967 and May 11, 1968 with the 2m - universal telescope of the Tautenburg Observatory /G.D.R./ and described in a separate paper /Bartl and Szumiejko,1975/.

Both the light-curves and the spectral development of the nova indicate that there are two separate and distinct phases of the outburst.

Phase I started with the outburst in July 1967 and lasted until December 1967. The spectrum in phase I shows lines which are normal for the early stages of a nova, and smoothly changing line profiles and velocities. The brightness also changes smoothly.

A sudden brightening of short duration of Nova Delphini in December 1967 was undoubtedly the beginning of its subsequent development phase. This event, generally interpreted as a second outburst, was the cause of substantial changes in the spectrum and in the behaviour of the light-curve. During this phase the light-curve varies rapidly and irregularly in the range of $1^m.2$ and as fast as $0^m.5$ during one night /Hutchings, 1970/. The spectrum shows also rapid changes, and each line possesses several absorption, and one strong, very broad emission component. A new series of lines, invisible up to that time appear in the spectrum.

All that which will be discussed below, occurs in the phase of the second outburst of the nova, i.e. in the interval from December 12, 1967 to May 13, 1968.

Fig. 1 shows the profiles of Hβ, He I 5876, Na I dublet, and O I 6158 lines. It can be seen that the position of Hβ absorption is not submitted to considerable changes in time, whereas He I line shows considerable radial velocity changes.

M. Friedjung (ed.), Novae and Related Stars, 87-90. All Rights Reserved.
Copyright © 1977 by D. Reidel Publishing Company, Dordrecht, Holland.

It also should be noted that the intensity of O I 6158 line
is changing: on the spectra on which the velocity of He I
line is the greatest, O I line is faint /line nearly disap-
pears/, while on the spectra with lower velocity of He I line
- O I line is strong.

The possibility of the periodicity of He I radial velocity has been explored on the base of 16 dates from the Tautenburg spectra and two additional dates obtained from spectra given kindly by Dr J.B.Hutchings and Dr Y.Yamashita to
authors disposal. By means of periodogram technics two possible periods has been established: 0.19 and 0.45, /Fig. 2/.
Unfortunately, the number of observations, still insufficient,
did not allow to choose one period only.

If the periodicity of He I radial velocity changes is
real, two models can then be proposed here:
1. He I 5876 is formed in the matter surrounding the
main component of the double system. This envelope takes part
in two motions: in the expansion with a velocity of about
1000 km/sec and in the rotation around the mass centre of the
double system with a velocity of the order 250 km/sec. This
would testify to the possibility to state the doubleness of
the nova also in the maximum of its light.
2. He I 5876 is formed in the matter surrounding two
components. This envelope is very active and the velocity of
its motion outside changes in it in the interval from 800 to
1300 km/sec. In favour of the second model speaks, perhaps,
Fig. 3b. In the rotary motion, the amplitude of the radial
velocity changes of O I, Mg II, Si II and Sc II lines ought
to be the same as the amplitude of He I radial velocity changes. Only the mean value of the velocity could change. In
other words, Fig. 3b could testify the existence of a stratification of the envelope in model 2. It may be that in favour
of this model speaks also Fig. 3a, showing the central depths
changes of He I and O I absorption lines in time of one of my
two "real" periods.

Summarizing, it can be stated that the above results
attest a great activity of Nova Delphini after its two main
explosions. On the ground of such a small number of observations, it is rather difficult to refer to a definite model.

The details of this investigation, as well as the results obtained from the new spectroscopic material kindly
thrown open to me by Prof. Ch. Fehrenbach at the Haute-Provence, will be published in Acta Astronomica.

REFERENCES

Bartl, E., and Szumiejko, E. 1975, Acta Astron., 25, 265.
Hutchings, J. B. 1970, Publ. Dom. Astrophys. Obs., 13, 347.

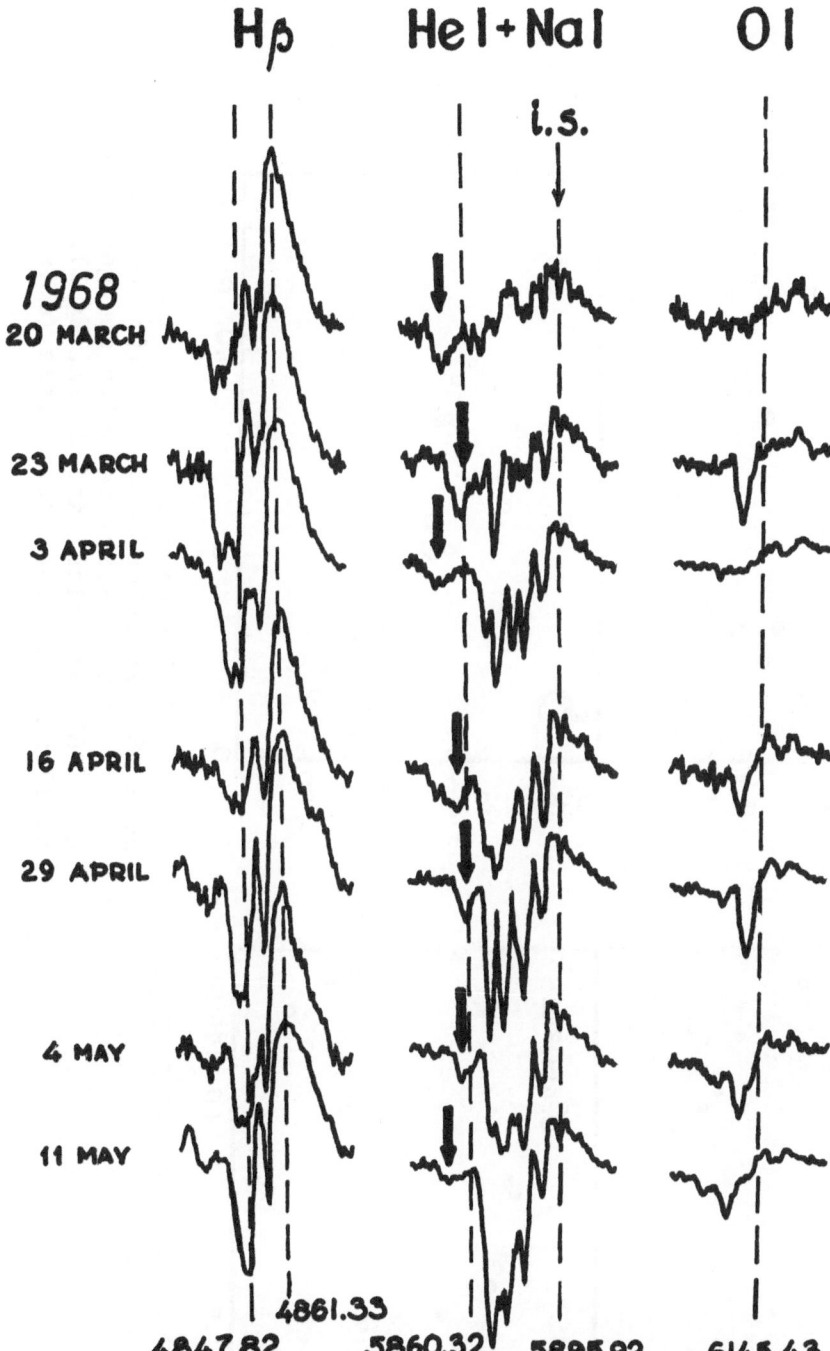

Fig. 1. The photographic darkening profiles in the second outburst phase of Nova Delphini 1967.

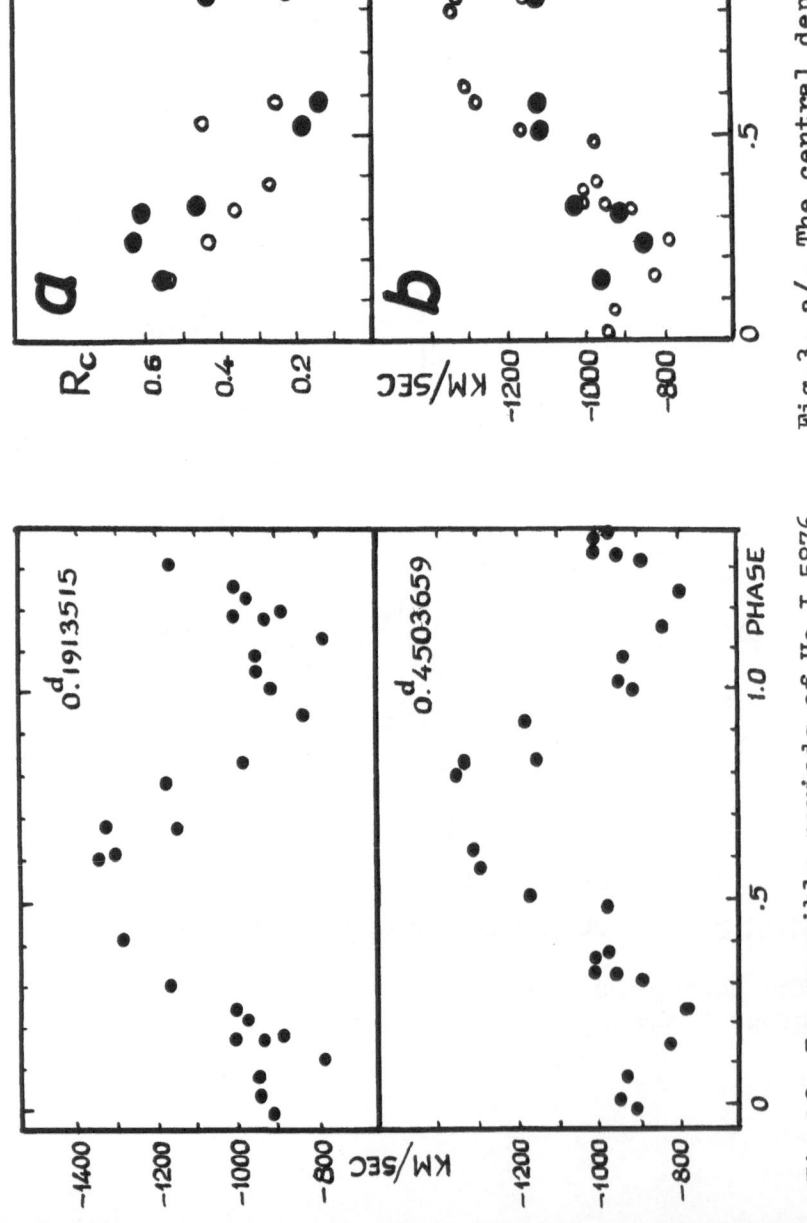

Fig.3. a/. The central depths changes of
He I 5876 /circles/ and O I 6158 /dots/.
       b/. The radial velocity changes of
He I 5876 /circles/ and Sc II, Mg II,
Si II and O I absorption lines /dots/.

Fig. 2. Two possible periods of He I 5876
radial velocity changes.

Model of Nova Envelopes

Kyoji Nariai

Tokyo Astronomical Observatory, Mitaka, Tokyo 181

The total energy of a nova phenomenon is estimated to be about 10^{45} ergs, which corresponds to the burning of 10^{26} grams of hydrogen. A part of the energy produced by nuclear reactions becomes the kinetic energy of the exploded shell, and the rest is lost from the system in the form of radiation. Before it is radiated away, the latter is stored for a certain period in the form of the potential energy. This has been demonstrated by the numerical calculations (Starrfield, Sparks, and Truran 1974), and by an analytical study (Nariai 1974). The latter predicts the increase in Te and almost constant L for the quasi-static envelope for a period of $0.5\,\tau_0$, where $\tau_0 = \kappa M_{env} / 4\pi cR_{core}$.

Based on the assumption of constant L, Nariai (1974) derived the position of the photosphere in the moving material. The result is a slowly decreasing radius of the photosphere in the initial phase and a sudden decrease.

References
Nariai, K. 1974 a, Publ. Astr. Soc. Japan, 26, 57.
Nariai, K. 1974 b, Astr. Astrophys., 36, 231.
Starrfield, S., Sparks, W. M., and Truran, J. W. 1974, Astrophys. J. Suppl. 28, 247.

M. Friedjung (ed.), Novae and Related Stars, 91. All Rights Reserved.
Copyright © 1977 by D. Reidel Publishing Company, Dordrecht, Holland.

"BLOATED DWARF" MODEL OF NOVA LIGHT CURVES

K. Brecher

Massachusetts Institute of Technology
Cambridge, Massachusetts, U.S.A.

ABSTRACT. A model is proposed to account for nova
light curves, especially those which show regular or
quasi-regular oscillations during decay. Various numerical calculations indicate that sudden nuclear
burning of matter accreted onto a white dwarf in a
binary star system can cause explosive ejection of
matter. However, if some of the matter ejected from
the white dwarf surface never escapes and tries to
return to the underlying star, a distended white dwarf
may temporarily result--a "Bloated Dwarf" (BD). The
continuum light curve would then come from two sources:
(1) the expanding ejected shell which becomes optically thin and weakly luminous after about one week
(but later producing most of the line spectrum) and
(2) radiation from the optically thick underlying BD.
With a white dwarf of mass $M_* \simeq M_\odot$, envelope mass $M_e \simeq$
10^{-4}-10^{-5} $M_\odot$ and radius $R_e \simeq 10^{11}$-10^{12} cm, the BD can
pulsate with periods of order 1-10 days. (By tran-
sit time arguments, an expanding shell is unlikely to
permit such variations.) Although the damping time of
the oscillations is difficult to estimate, one finds
that contraction of the envelope will occur on a time
scale of months. The increase of temperature with decreasing visual luminosity observed in several novae
arises from the shift in the peak emissivity toward
the UV as the BD contracts. The model also accounts
for the outburst size-decay time relation; the continuing activity, including mass loss, long after the initial event; and the existence of strong circular po-
larization in the continuum light.

M. Friedjung (ed.), Novae and Related Stars, 92. All Rights Reserved.
Copyright © 1977 by D. Reidel Publishing Company, Dordrecht, Holland.

PHYSICAL PARAMETERS OF PRENOVAE AND OUTBURST AMPLITUDES

Waltraut Carola Seitter

Astronomisches Institut der Universitaet Muenster, FRG

The best observed novae of C. Payne-Gaposchkin's list and the brightest novae found since 1960 yield 37 data points in a diagram showing absolute magnitudes of novae at minimum light versus amplitude of the outburst. Three dwarf novae are found well separated from the novae, five recurrent novae mix with the high luminosity end of the nova sequence. Due to the small scatter in absolute magnitudes at maximum a noticeable dependence of outburst amplitude on M_{min}, which ranges from 9^m to brighter than 0^m, is found. For eleven novae blue spectra at minimum indicate high temperatures. O-type temperatures and white dwarf densities are, however, mutually exclusive over a wide range of M_{min}. Only two novae, CP Pup and V1500 Cyg, have, under the assumption of O-type temperatures, densities within the white dwarf range and even they are found near the low-density limit. This leaves three explanations for minimum light:
1) All prenovae are normal white dwarfs, additional light comes from a sufficiently bright blue disk.
2) All prenovae are white dwarfs with a much wider range of temperatures than hitherto assumed - up to more than a million degrees.
3) All prenovae are O-type stars with a wide range of densities. These assumptions lead, respectively, to a: 1) disk-amplitude relation; 2) temperature-amplitude relation; 3) density-amplitude relation; stating that fainter disks, lower temperatures or higher densities agree with more spectacular nova outbursts.

The above findings locate prenovae in the HR-diagram of evolved stars to the right of the nuclei of planetary nebulae in case 3), to the left in case 2). Case 1) suggests that larger disks substitute for more violent outbursts. In all cases, the nova event appears as a way to circumvent the state of a planetary nebula.

M. Friedjung (ed.), Novae and Related Stars, 93. All Rights Reserved.
Copyright © 1977 by D. Reidel Publishing Company, Dordrecht, Holland.

TIMING OF THE SHELL EJECTION IN NOVA OUTBURSTS

Elia M. Leibowitz

Department of Physics and Astronomy
and the Wise Observatory
Tel-Aviv University
Tel-Aviv, Israel

Some time after the outburst of a nova in the visible light its light intensity is dominated by the flux in the emission lines. The rate of fading of the nova in such times is the rate of weakening of the emission lines. The intensity of a recombination or a forbidden line emitted by a given mass of ionized gas is proportional to the density. Thus the light curve gives the expansion rate of the nova shell. For a shell with a constant expansion velocity the light curve therefore enables the determination of the beginning of the expansion process, namely the timing of the ejection of the shell. In four classical Galactic novae the ejection of the main shell is found to occur between 19 and 40 days before the outburst in the visible light.

M. Friedjung (ed.), Novae and Related Stars, 94. All Rights Reserved.
Copyright © 1977 by D. Reidel Publishing Company, Dordrecht, Holland.

NOVA FH SERPENTIS AS A TEST OF OUTBURST MODEL

M. Friedjung

Institut d'Astrophysique, Paris
98 bis, Bd Arago
75014 Paris, France

FH Serpentis is one of only two novae observed in the UV, IR, and visible for which new constraints on nova models can be obtained. The UV measurements of Gallagher and Code (Ap. J. 189, 303 1974), combined with the IR observations of Geisel et al (Ap. J. 161, L 101, 1970), indicate that if the interstellar absorption correction of Gallagher and Code is correct, the energy distribution was nearly Planckian for more than a month after the visual maximum. Photospheric temperatures and radii can be determined, as well as luminosities taking into account radiation of all wavelengths. In the first month the temperature determined rises from 5200 to 9200°, the radius decreases from 320 to 60 $R_\odot$, and the luminosity (near the Eddington limit) decreases from 6.9 to 2.4 x 10^4 $L_\odot$. Thus the last quantity may not have been constant as claimed by Gallagher and Code. These results are unfortunately sensitive to the corrections for interstellar absorption; if the true correction is 1 1/2 times larger, the energy distribution is less Planckian, the temperatures higher, and the luminosity almost constant.

The photosphere cannot be formed by stationary gas, as a comparison with supergiants of similar temperature and expected surface gravity, would lead one to expect strong unshifted lines, not found on a Haute Provence Observatory plate. Thus the photosphere was probably produced by continuously ejected gas.

At a temperature of 5200°, most hydrogen would be neutral, and the photospheric opacity small. Most mass could have been ejected near the visual maximum.

Spectroscopy of the Nova Vulpeculae 1968 No.1 (LV Vul)

M. Sobotka, S.Štefl; Department of Astronomy and Astrophysics, Faculty of Mathematics and Physics, Charles University, Praha, and
J. Grygar; Astronomical Institute of the Czechoslovak Academy of Sciences, Ondřejov; Czechoslovakia

Nova LV Vul was observed with the 2m telescope in Ondřejov from April 28 to October 15, 1968. 13 spectrograms in the region 360 – 505 nm were obtained with dispersions from 1.6 to 7.5 nm mm^{-1}. Some 90 emission lines were identified (H I, He I, He II, O II, O III, /O II/, /O III/, N II, N III, Fe I, Fe II, Fe III, /Fe II/, /Fe III/, /Fe V/, C II, C III, Ti II, Cr I, Cr II, V II). Three emission peaks were found in the profiles of broad hydrogen emissions, giving the radial velocities of (-780 ± 30) km s^{-1}, (-160 ± 50) km s^{-1} and $(+650\pm40)$ km s^{-1}. From measurements of the emission halfwidths we have obtained the average expansion velocity of (1700 ± 300) km s^{-1}. Dorschner et al. (AN **291**, 1969, 217) derived the absolute visual magnitude in maximum M = $= (-7.5^m \pm 0.5^m)$, and the distance r = (1450 ± 400) pc.

Then, if we assume the colour temperature near maximum close to 10000 K, the mass of the ejected envelope is (1.4 ± 0.2) x 10^{-5} solar masses. The nova radius near maximum was R = (110 ± 15) $R_\odot$. From the combination of the photometric and spectroscopic data we can derive the time when the angular diameter of the envelope will reach 3″. Under some simplified assumptions we infer that the envelope had this diameter some 2250 days after maximum light, i.e. in mid-June 1974. The corresponding linear diameter was 2200 a.u. = 4.7 x 10^5 $R_\odot$. Thus, there is a fair chance to detect the envelope by direct photography right now, since the nebular diameter must be close to 4″ .

GRAIN FORMATION IN NOVA ENVELOPES

T. YAMAMOTO and S. NISHIDA *
DEPARTMENT OF PHYSICS , KYOTO, JAPAN
*DEPARTMENT OF INDUSTRIAL MANAGEMENT, SETSUNAN UNIVER-
SITY, NEYAGAWA, OSAKA,JAPAN

Grain formation in ejected envelopes of novae is investigated. The solar abundance is assumed as one of the typical examples
that C/O is smaller than unity. Then, magnesium silicate is proved
to be the first main condensate. We estimate the number density
$c_1(t_1)$ of monomers (SiO molecules for our case) at the beginning
of condensation to be about 10^5 cm for $\Delta M = 3 \times 10^{-4}$ $M_\odot$ and
V_{exp} = 1000 km sec^{-1} (ΔM is the ejected mass and V_{exp} is the
expansion velocity).

Next, we calculate the time variation of the number density
of magnesium-silicate grains of various sizes, based on the non-
steady state nucleation theory. The size distributions are found
to be flat and have no conspicuous peak. Particle sizes are
widely distributed from, say, several hundred Å to the order of
Å. The lower limit of the ejected mass for forming grains is about
10^{-4} $M_\odot$. If the ejected mass is larger than this value, almost
all the monomers condense into grains within about ten days.

The maximum radius of grains a_{max} is estimated to be about
50 Å for $c_1(t_1) = 5 \times 10^5$ cm^{-3}. The infrared luminosity is
also calculated and is found that these grains can radiate
several x 10^4 $L_\odot$. This is sufficient to explain the infrared
excesses of novae.

M. Friedjung (ed.), Novae and Related Stars, 97. All Rights Reserved.
Copyright © 1977 by D. Reidel Publishing Company, Dordrecht, Holland.

PART III

<u>(a)   NEBULAR STAGE OF NOVAE</u>

PROBLEMS OF IONIZATION OF NOVA ENVELOPES
IN THE NEBULAR STAGE. THE MODELS FOR NOVA DELPHINI 1967

Romuald TYLENDA

Institute of Astronomy, Nicolaus Copernicus University,
Toruń, Poland.

ABSTRACT

The analysis of spectra of novae in the nebular stage shows that the absorption of ultraviolet radiation produced by a central source is the main mechanism of ionization of nova envelopes. The ionization and thermal equilibria are generally satisfied except the late nebular and post-nova stages when the recombination time is longer then one year. In the case when [OI] lines are observed in the spectrum of a nova, the envelope is expected to be optically thick in the Lyman continuum. Due to large overabundances of CNO elements observed in novae, the absorption of ionizing photons by these elements is important and must be taken into account in model calculations.

A sequence of stationary photoionization models of the envelope of Nova Del 1967 in the nebular stage have been constructed. The results of calculations have been compared with spectroscopic observations of the nova made in June, 1969 - August, 1975. The observed intensities of emission lines cannot be entirely matched by the models of the envelope in which a single blackbody spectrum of the ionizing radiation is assumed; the computed intensities of [NeV], [OII], and [NII] lines are then 4-100 times smaller then the observed ones. After detailed discussion we have come to the conclusion that the ionizing radiation is probably produced by two sources with

M. Friedjung (ed.), Novae and Related Stars, 101-122. All Rights Reserved.
Copyright © 1977 by D. Reidel Publishing Company, Dordrecht, Holland.

different temperatures. The observational data obtained in June-July, 1969 have been well matched by the
models assuming that the ionizing radiation is a sum
of two blackbody distributions with temperatures:
$\sim$2.5 $\times$ 10^5 K and $\sim$4 $\times$ 10^4 K. The hotter spectrum is
attributed to the hot atmosphere of the primary where-
as the cooler one is probably produced by the heated
secondary and/or the outburst remnent transformed into a differentially rotating disk.

Finally, we discuss briefly the evolution of the
envelope of Nova Del 1967 in the nebular and post-nova
stages.

1. INTRODUCTION

Intensities of emission lines observed in spectrum
of a nova in the nebular stage depend on the relative
numbers of atoms in various stages of ionization as
well as on the electron temperature and density in a
nova envelope. On the other hand, the physical state
of the matter in the envelope is determined by mechanisms of ionization, heating, and cooling of the gas.
Therefore a comprehensive and self-consistent analysis
of the spectrum of a nova should take into account
not only the mechanisms of line emissions but also
the processes controlling the physical conditions of
the gas. This can be done only by theoretical calculations of a model envelope giving the run of the
electron temperature, ionization structure, and emission coefficients with the radius. The predicted intensities of emission lines are then compared with
the observed values and information about the source
of ionization as well as the element abundances, and
the density in the envelope are obtained.

In this paper we summarize the basic problems of

calculations of ionization structure models of nova
envelopes. Next we present the main results of the
model analysis of Nova Del 1967 in the nebular stage.

2. THE CALCULATION OF THE MODEL ENVELOPE

A calculation of a theoretical model of a nova
envelope as well as other line-emission regions must
be preceded by a careful discussion of possible sources of ionization, heating, and cooling of the gas,
and their role in a considered case. We must also
decide whether stationary models can be constructed
or rather time-dependent calculations are necessary.
It can be done by means of estimations of recombination
and cooling times in the envelope. Determination of a
'transparency' of the gas to the ionizing agent is also
important when an ionization structure model is calculated. When the envelope is opaque we have the ioniza-
tion-bounded case and calculations are performed up
to the exhaustion of the ionizing agent by the gas.
Otherwise, the envelope is density-bounded and its
geometrical thickness must be assumed or found from
other considerations.

In the first part of this section we discuss the
above mentioned problems in more detail.

2a. Ionization and heating mechanisms.

The electron temperature obtained from intensity
ratios of forbidden lines in spectra of novae in the
nebular stage is close to 10^4 K (e.g. Pottasch 1959b;
Malakpur 1973). This as well as strong Bowen's lines

observed in nova spectra (e.g. Tylenda 1977) strongly
support that the absorption of ultraviolet photons by
the gas is the dominant mechanism of ionization and
heating of nova envelopes in the nebular stage. On
the other hand, there are observational arguments that
other processes of energy injection into the gas, such
as collisions with thermal or suprathermal electrons,
or with suprathermal protons, are negligible (cf.
Tylenda 1977). Ultraviolet observations of Nova Ser
1970 by Gallagher and Code (1974) have shown that long
after the visual maximum of brightness the bolometric
luminosity of the nova remains almost constant but the
effective temperature continuously rises. This has
also been confirmed by theoretical investigations of
Starrfield et al.(1974a,b) and Nariai (1974). Therefore the central star can be expected to radiate a
large amount of extreme-ultraviolet photons ($\lambda \leqslant 912$ Å)
during the nebular stage.

2b. Cooling mechanisms.

    Free electrons lose their kinetic energy in re-
combination, free-free radiation, and first of all in
collisional excitation of lowest levels of different
ions.

    In the late nebular and post-nova stages, when
$N_e \lesssim 10^5$ cm^{-3}, the energy loss processes are similar
to those in planetary nebulae (see e.g. Harrington
1968; Osterbrock 1974). However, in the early nebular
stage, when $N_e \gtrsim 10^6$ cm^{-3}, all fine-structure transitions are strongly suppressed by collisional de-exci-
tation. Most of forbidden lines, such as [OIII] lines,
are also much less important coolers than in planetary

nebulae or HII regions. Therefore allowed and semi-
forbidden lines, such as CIV λ 1549 and CIII] λ 1909
lines, come into prominence as in the case of quasar
emission regions (see e.g. Davidson 1972,1973).

Other sources of cooling, such as adiabatic expansion of a nova envelope, are completely negligible
in the nebular stage.

2c. Recombination and cooling times.

The characteristic time of recombination in an
ionized hydrogen gas is approximately

$$t_{rec} \cong \frac{1}{N_e \, \alpha_B(T_e)} \cong \frac{4.5 \times 10^7}{N_e} \text{ days},$$

where α_B is a recombination coefficient upon all levels of hydrogen except the ground one. The numerical
value is given for $T_e = 10^4$ K. The cooling time is
usually shorter than t_{rec} (cf. Gallagher and Anderson
1976). At the beginning of the nebular stage $N_e \gtrsim 10^7$
cm^{-3} and $t_{rec} \lesssim 4.5$ days so the recombination proceeds
much faster than the evolution of a nova envelope.
Therefore the assumption of ionization and thermal
equilibria is then completely justified and stationary
models of the envelope can be constructed. However,
as the envelope expands, the density quickly decreases
and in the post-nova stage, when $N_e \cong 10^4$ cm^{-3}, t_{rec}
$\cong 10$ years. Thus rather time-dependent models are
needed to study the evolution of a nova envelope in
the final nebular and post-nova stages (cf. Mustel
and Boyarchuk 1970; Gallagher and Anderson 1976).

In this paper we discuss stationary models only.
Few time-dependent calculations have recently been

performed for the interstellar medium (e.g. Gerola et
al.1973; Schwarz 1973) and an old planetary nebula
around FG Sge (Harrington and Marionni 1976). There
are no such models available for the nova envelope case.

2d. Optical thickness of a nova envelope.

The optical thickness of an envelope to the H-
ionizing radiation can be estimated from an analysis
of [OI] lines in a nova spectrum. The ionization potentials of O and H are nearly the same and a charge-
exchange reaction between ions and neutral atoms of H
and O is very effective (Field and Steigman 1971).
Therefore, if [OI] lines are observed in the spectrum
of a nova, they indicate the presence of neutral H,
and thus the envelope is expected to be optically
thick to H-ionizing photons (cf. Osterbrock 1974).
The [OI] lines are very often observed in spectra of
novae (e.g. Payne-Gaposchkin and Gaposchkin 1942;
Andrillat and Houziaux 1970a,b,1971; Andrillat and
Collin-Souffrin 1974). Rough estimations of the optical thickness of the envelope of Nova Del 1967 in
June-July,1969 give the value $\gg 5$ (Tylenda 1977).
Therefore it seems that nova envelopes are usually
optically thick to the H-ionizing radiation in the
nebular stage although it may be not the case in the
post-nova stage (Tylenda 1977).

2e. Computational method.

Details of calculations of an ionization structure model of a gas shell or cloud excited by a central
source of radiation are presented in a number of papers

devoted to model planetary nabulae (e.g. Harrington
1968; Flower 1969) or model emission regions of quasars
(e.g. Davidson 1972; MacAlpine 1972) as well as in the
excellent book of Osterbrock (1974). Therefore only
the general outline of the computational method used
for calculations of models of the envelope of Nova Del
1967 is described here. Some more details including
sources of atomic parameters will be presented in a
paper which is now in preparation (Tylenda 1977).

In our calculations the spherically symmetric
envelope[x) with a constant density has been assumed
to be optically thick to the primary ionizing continuum produced by a small central source. The HI, HeI,
and HeII Lyman continua, and HeII L$_\alpha$ diffuse emission
have been treated in the on-the-spot approximation.
The elements H, He, C, N, O, and Ne have been included.
The ionization has been calculated by solving the
steady-state equations including photoionization, radiative recombination, and O-H charge transfer. The
electron temperature has been found by solving the
energy balance equation. Electrons have been assumed
to gain their energy by photoionization of all ions.
The recombination of H and He ions, free-free transitions, and collisional excitations of permitted, semi-
forbidden, and forbidden lowest transitions have been
included into the energy loss. The allowance for col-

[x)] It is known from observations that the gas is not
distributed spherically in nova envelopes but rather
in form of equatorial rings and polar blobs (Mustel
and Boyarchuk 1970; Hutchings 1972). Therefore a
filling factor, ξ, has been introduced in our calculations. We have adopted $\xi = 0.5$ throughout. The results have appeared to be rather insensitive to the
value of ξ (see Tylenda 1977).

lisional de-excitations of metastable states has been
made.

Various determinations of chemical composition of
nova envelopes give large overabundances of CNO elements
(Pottasch 1959c; Mustel and Baranova 1965; Antipova
1974). The calculations show that, due to those overabundances, the electron temperature does not exceed
the value of 2×10^4 K in any region of the nova envelope. Therefore collisional ionization and dielectronic
recombination are negligible compared with photoionization and radiative recombination respectively. On the
other hand, the absorption of ionizing photons by CNO
elements becomes imporatant in the nova envelope and
so the contributions of all ions to the optical depth
of the ionized gas must be taken into account in model
calculations.

The emission line luminosities produced by the
model envelope have been found from volume integrals
of respective line-emission coefficients over the ionized region.

3. NOVA DELPHINI 1967 IN THE NEBULAR STAGE

3a. Observational data.

The observational material we have used consists
of three sets of emission line luminosities obtained
during three observational seasons.

The essential and reachest material was collected
by Dr A. Woszczyk at McDonald Observatory in June and
July,1969. The spectra were taken at dispersion of 17
Å/mm and covered the wide range of wavelengths: from

3400 Å up to 7400 Å. The results of measurements of
line intensities made on these spectra have been used
as the main test of correctness of constructed models.

The second set of observational data is that obtained by Andrillat, Fehrenbach, and Houziaux (1974)
in 1971-72.

The last series of spectra used here was performed
by Dr A. Woszczyk at Haute-Provence Observatory in
August,1975 when Nova Del was already in its post-nova
stage.

The details of reduction of spectra and the full
sets of observational results will be published later
(Tylenda 1977).

3b. Models for the first observational season: June-
    July,1969.

Physical conditions in a model envelope are fully
determined by chemical composition, number density,
and character (spectral distribution) of the ionizing
radiation. Absolute dimensions of the ionized part of
the envelope depend on the total amount of ionizing
photons.

First, we have tried to construct a model of the
Nova Del envelope assuming that the ionizing radiation
is produced by a single star radiating as a blackbody.

Models with a single blackbody spectrum of the
ionizing radiation. A model has been fitted to the
observations as follows. The temperature of the ionizing star has been chosen to fit the observed intensity
of the HeIIλ4686 line. The luminosity of the star has

been found to match the observed luminosity in the H_β
line. The number density of hydrogen atoms has been
adjusted to find the observed ratio of [OIII] λ 4363
to [OIII] $\lambda\lambda$ 4959 and 5007 lines. Abundances of He, O,
and Ne have been chosen to fit the observed intensities
of HeI, [OIII], and [NeIII] lines respectively. Nitrogen has been assumed to be as abundant as oxygen. The
abundance of carbon has been taken to be equal either
to the cosmic value (Allen 1973) or to the value found
by Mustel and Baranova (1965) for Nova Her 1934.

It has appeared, however, that if the observed
strengths of the above mentioned lines have been matched by the model, then the other lines have been calculated to be considerably weaker than observed; the
computed intensity of the [NeV] λ 3426 line is four
times smaller than the observed one whereas the [OII]
and [NII] lines are down by a factor of 100.

The similar discrepancy between the observations
and the calculations for the lines resulting from lower
ionization stages (e.g. [OII] and [NII]), although to
a less degree, also appeares in the case of planetary
nebulae (e.g. Harrington 1968,1969). Several authors
propose to take density condensations (Harrington 1969;
Williams 1970) and/or dust (Balick 1975) into account
in theoretical models to enhance the emission from the
lowest stages.

However, the disaccordance between the calculations and the observations in the case of Nova Del
cannot be removed by the density inhomogeneities or
the internal dust introduced into the model envelope.
Calculations with the density condensations included
in the form proposed by Harrington (1969) have given
the increase of the [OII] and [NII] lines by a factor

of 2-3 only. A model in which the dust has been mixed
with the gas, similar to those calculated by Balick
(1975), has not been able to increase the intensities
of [OII] and [NII] lines sufficiently, either. It is
necessary to add that infrared observations made by
Geisel et al.(1970) evidence that the dust really
exists in the envelope of Nova Del but its amount is
rather small and therefore it may be neglected in the
calculations of the ionization structure[x] (cf. Tylenda 1977).

The discussion of the ionization structure of the
model envelope suggests that a better match with the
observational data will be obtained if the ionizing
radiation is assumed to be a sum of two distributions
with different temperatures (for more detail see Tylenda 1977). This conclusion is also cofirmed by the
following qualitative analysis of interaction between
a rotating binary system and a remnant of the outburst
surrounding both components of the system (see also
Tylenda 1976,1977).

Novae are known to be close binary systems (e.g.
Kraft 1964). The secondary component of such a system
fills its Roche lobe and transfers the hydrogen-rich
matter through the inner Lagrangian point onto the
primary which is a white dwarf. According to Starrfield, Sparks, and Truran (1974a,b) a nova outburst
is due to unstable thermonuclear runaways in a hydrogen-rich envelope of the carbon-oxygen white dwarf.

---

[x] The role of the dust in envelopes of some novae,
such as Nova Her 1934 or Nova Ser 1970, is very
important (Geisel et al.1970; Clayton and Hoyle
1976) and therefore it can influence significantly
the ionization structure there.

Part of the hydrogen-rich envelope escapes from the
system and is observed as the expanding envelope whereas the rest forms an extended remnant. After the outburst the white dwarf is still burning hydrogen in the
shell source although at a much smaller rate than during
the outburst. In the remnant the energy is transported
by convection.

The radius of the remnant exceeds the semimajor
axis of the underlying system and so a common envelope
for the binary is produced. Then a drag force should
be experienced by the rapidly rotating binary, particularly by the secondary. Friction due to the drag
generates heat. Following Paczyński (1976) the drag
luminosity, L_D, may be estimated from the formula:

$$L_D \cong \pi R_2^{\ 2} v_r^{\ 3} \varrho ,\qquad\qquad (1)$$

where R_2 is a radius of the secondary, v_r - orbital
velocity of the secondary, and ϱ - density of circumbinary matter. For the 10^{30} g remnant extended to
$\sim 10^{11}$ cm (Starrfield et al. 1974a,b) we get $\varrho \cong 10^{-4}$
g cm^{-3}. Thus for typical parameters of a nova binary
system formula (1) gives $L_D \cong 10^{39}$-10^{40} ergs s^{-1}.
This value exceeds significantly the critical Eddington
luminosity for a 1 $M_\odot$ star. Part of that energy produces a continuous mass ejection and a rest of it heats
the secondary. Friedjung (1975) has suggested that a
certain portion of convective energy of the remnant
can also be transmitted to the secondary causing its
additional heating.

The drag luminosity can be significantly reduced
when the circumbinary matter begins to co-rotate with
the system. On the other hand, the convection should

transport the energy and the angular momentum outwards.
This should accelerate the mass ejection as well as
slow down the rotation of the matter in close vicinity
of the binary.

After some time, when enough angular momentum is
transmitted from the system to the circumbinary matter,
the remnant should be transformed into a differentially
rotating disk around the system. If so then physical
processes occuring in the remnant resemble those in
accretion disks around compact objects in binary systems (see e.g. Shakura and Sunyaev 1973; Lynden-Bell
and Pringle 1974). Strong heating of inner parts of
such a disk takes place and a considerable amount of
ionizing photons can be radiated.

According to recent investigations of mass outflow
from novae by Bath and Shaviv (1976) the transition
stage in optical light curves occurs when the remnant
has dimensions comparable with the underlying binary
separation. Thus we suggest that the nebular stage
begins when both components of the system become visible
and the hot radiation produced by them can reach the
expanding envelope. We note that due to the nuclear
burning lasting long after the outburst in the deepest
layers of the hydrogen-rich remnant (Starrfield et al.
1974a,b) the surface of the primary should be very hot
(probably much hotter than 10^5 K) in the nebular stage,
at least at the beginning of it.

Concluding, the ionizing radiation in the nebular
stage can be represented, in a first approximation,
as a sum of two constituents: one - hotter - is attributed to the hot atmosphere of the primary, the other
- cooler - is produced by the heated secondary and/or

the outburst remnant transformed into a differentially
rotating disk.

Finally, we mention that in the post-nova stage
the ionizing radiation is also expected to consists of
two components; the cooler radiation is produced by an
accretion disk formed around the white dwarf while the
hotter one is emitted by a boundary layer between the
disk and the atmosphere of the primary.

Table 1

Relative intensities of emission lines
for Nova Del 1967 in June-July,1969. ($I(H_\beta)$ = 1.00)

Lines		Model 2S-A-I	Model 2S-B-I	Observations
HeI	4471	0.075	0.079	0.073
	5876	0.20	0.21	0.38
HeII	4686	0.264	0.256	0.266
[NII]	5755	1.06	1.08	1.07
	6548	0.21	0.21	0.24
	6583	0.62	0.61	0.86
[OI]	6300	0.018	0.027	0.15
[OII]	3728	0.045	0.047	0.045
	7319	0.71	0.75	0.7
	7330	0.54	0.57	0.7
[OIII]	4363	1.32	1.28	1.33
	4959	1.83	1.88	1.89
	5007	5.26	5.41	5.2
[NeIII]	3869	0.61	0.63	0.65
[NeIV]	4720	0.016	0.016	$\leqslant$ 0.03
[NeV]	3426	0.48	0.49	0.47
$L(H_\beta)$[ergs/s]		6.2×10^{34}	6.3×10^{34}	6.3×10^{34}

Table 2

Input parameters of the models.

	Model 2S-A-I	Model 2S-B-I
Central source		
T_1 [K]	2.5×10^5	2.5×10^5
L_1 [ergs s^{-1}]	8.52×10^{36}	9.02×10^{36}
T_2 [K]	3.5×10^4	4.0×10^4
L_2 [ergs s^{-1}]	1.95×10^{37}	1.65×10^{37}
Envelope		
N(H) [cm^{-3}]	5.3×10^6	6.7×10^6
He/H	0.27	0.27
C/H	4.0×10^{-4}	1.7×10^{-3}
N/H	3.5×10^{-3}	5.4×10^{-3}
O/H	4.8×10^{-3}	8.5×10^{-3}
Ne/H	2.4×10^{-4}	4.0×10^{-4}
M_{env} [g]	2.9×10^{29}	2.1×10^{29}

Models with the ionizing radiation being a sum of two blackbody distributions. Following the above discussion we have tried to construct a model of the Nova Del envelope assuming that the ionizing radiation is a sum of two blackbody distributions with different temperatures. Relative intensities of emission lines calculated in two such models together with the observed values (corrected for the reddening) are presented in Table 1. The input parameters of the models are listed in Table 2. The abundances of the elements are in terms of the number of atoms relative to the number of hydrogen atoms. The mass of the envelope,

M_{env}, given in the last line of Table 2, is the mass of the ionized part of the envelope.

Because of lack of suitable carbon lines in the observed range of wavelengths the carbon abundance cannot be determined by matching the model results to the observations. Therefore we have performed the calculations with two different values of the carbon abundance; model 2S-A-I has the abundance close to the cosmic value (Allen 1973) whilst in model 2S-B-I the value found by Mustel and Baranova (1965) for Nova Her 1934 has been adopted.

It can be seen from Table 1 that the agreement between the observational data and the calculated models is quite good except the [OI]λ6300 line. However, this line is produced in a narrow transition region between the ionized and neutral parts of the envelope where dynamical effects may be important (Mallik 1975).

Figure 1 presents the resultant distribution of the electron temperature and density. The ionization structure of H, He, and O in model 2S-A-I is shown in Figure 2.

We state the considerable overabundances of N and O in the envelope of Nova Del 1967; the overabundance of N is about 50 times and that of O - 10 times compared with cosmic abundances (Allen 1973). Helium also appears to be about 3 times enhanced in the Nova Del envelope. The CNO overabundances found for Nova Del 1967 as well as for other novae by various authors (Pottasch 1959c; Mustel and Baranova 1965; Antipova 1974) are in good agreement with hydrodynamic models of the nova outburst studied by Starrfield, Sparks, and Truran (1974a,b).

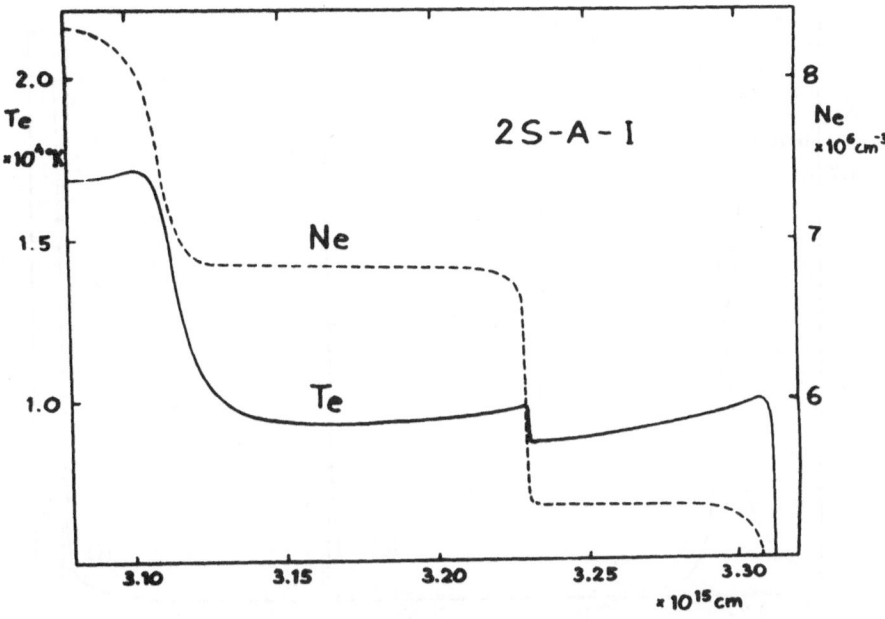

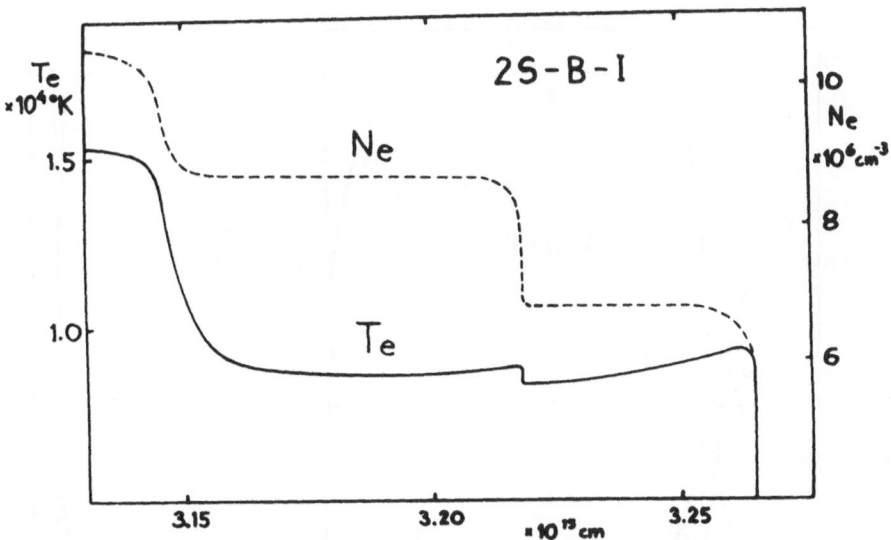

Fig. 1. The calculated distributions of the electron
temperature and density in models 2S-A-I (top)
and 2S-B-I (bottom).

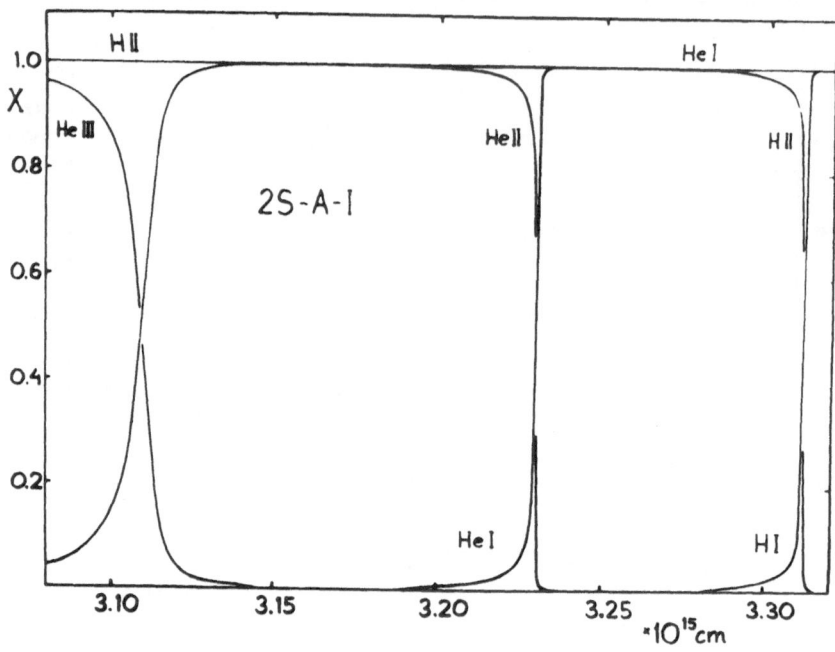

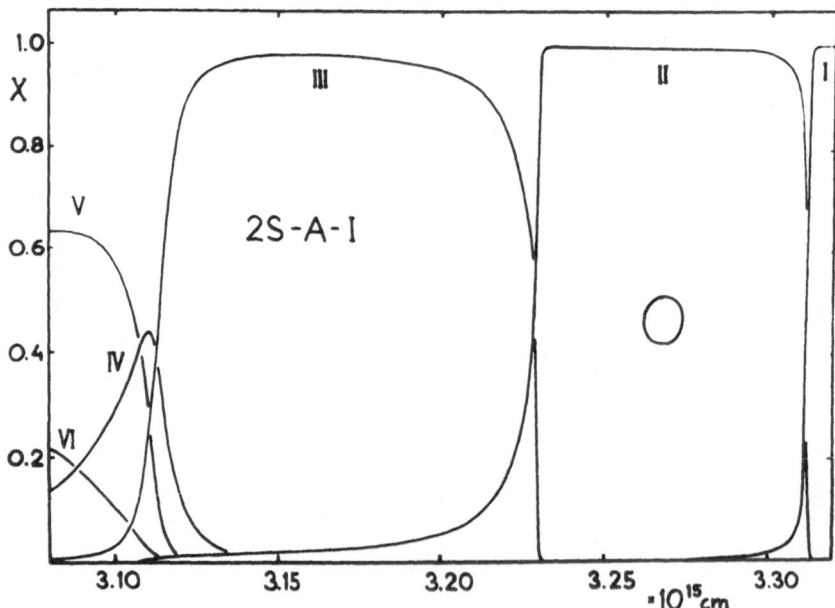

Fig. 2. The ionization structure of model 2S-A-I:
hydrogen and helium (top), and oxygen (bottom).

3c. Evolution of Nova Del 1967 in the nebular stage.
    Models for 1971-72 and August,1975.

Let us now describe briefly some results of the
model analysis of evolutionary changes in the nebular
spectrum of Nova Del 1967. The observations we have
used were made in 1971-72 (Andrillat et al.1974) and
August,1975. We have tried to match these observations
by the above presented models, 2S-A-I and 2S-B-I, constructed for June-July, 1969, taking into account
changes due to expansion of the envelope and an unknown
evolution of the central source of the ionizing radiation. The chemical composition of the envelope has been
assumed to remain unchanged during the nova evolution.

The expansion increases the radius of the enevlope
and decreases its density. The density of hydrogen
atoms in the envelope appeared to be ~2.5-3.0 × 10^5
cm^{-3} in 1971-72 and ~2.5 × 10^4 cm^{-3} in August,1975
compared with ~6 × 10^6 cm^{-3} in June-July,1969. That
decrease is somewhat faster than resulting from the
adiabatic expansion of a gas shell into vacuum (e.g.
Pottasch 1959a). Radiation pressure from the absorption
of the ionizing continuum and the scattering of reso-
nance lines created in the envelope can be important,
as in quasar clouds (e.g. McKee and Tarter 1975), and
can cause the faster expansion of the nova envelope.

The luminosity of the central system quickly decreased with time. It appeared to be ~2.5 × 10^{36} ergs
s^{-1} in 1971-72 compared with the value of ~2.8 × 10^{37}
ergs s^{-1} in June-July,1969. The temperature of the
hotter constituent was not changing appreciably and
remained ~2.5-3.0 × 10^5 K in 1971-72 but that of the
cooler source was probably diminishing slightly with

time and it was $\sim 2.5-3.0 \times 10^4$ K in 1971-72.

In August, 1975 Nova Del was already in its post-nova stage. The ionizing radiation was then probably produced by an accretion disk around the primary and a boundary layer formed between the disk and the atmosphere of the primary. The lack of [OI] lines in the nova spectrum suggested that the envelope was at least partially transparent to the H-ionizing photons although it still remained opaque to the HeII-ionizing radiation. Therefore the results of the calculations appeared to be sensitive only to the parameters of the hotter constituent (L_1, T_1) that means of the boundary layer which produced mainly the HeII-ionizing photons. The luminosity of the boundary layer was close to 8×10^{34} ergs s^{-1} and its temperature $\sim 2-3 \times 10^5$ K. Estimations show that if the primary is a 1 $M_\odot$ white dwarf then a mass accretion rate $\sim 3 \times 10^{-8}$ $M_\odot$/year is required to explain the derived luminosity and temperature of the boundary layer. This value is in good agreement with mass transfer rates estimated for other post-novae.

Finally, we would like to note that, as it has been mentioned in section 2c, the recombination time in the nova envelope during the post-nova stage is of the order of years. Therefore the above results based on the stationary models should be taken as approximative only. Time-dependent calculations are necessary to study more accurately the final nebular and post-nova stages of the nova.

REFERENCES

Allen,C.W.,1973,'Astrophys.Quantities',third ed.,London.

Andrillat,Y.and Collin-Souffrin,S.,1974,Astron.Astro-
        phys.31,347.
Andrillat,Y.,Fehrenbach,C.,and Houziaux,L.,1974,
        Astrophys.Space Sci.31,169.
Andrillat,Y.nad Houziaux,L.,1970a,Astrophys.Space
        Sci.6,36.
Andrillat,Y.and Houziaux,L.,1970b,ibid.9,410.
Andrillat,Y.and Houziaux,L.,1971,ibid.13,100.
Antipova,L.I.,1974,Highlights of Astronomy 3,501.
Balick,B.,1975,Astrophys.J.201,705.
Bath,G.T.and Shaviv,G.,1976,Monthly Notices Roy.Astr.
        Soc.175,305.
Clayton,D.D.and Hoyle,F.,1976,Astrophys.J.203,490.
Davidson,K.,1972,Astrophys.J.171,213.
Davidson,K.,1973,ibid.181,1.
Field,G.B.and Steigman,G.,1971,Astrophys.J.166,59.
Flower,D.R.,1969,Monthly Notices Roy.Astr.Soc.146,171.
Friedjung,M.,1975,private discussions.
Gallagher,J.S.and Anderson,C.M.,1976,Astrophys.J.
        203,625.
Gallagher,J.S.and Code,A.D.,1974,Astrophys.J.189,303.
Geisel,S.L.,Kleinmann,D.E.and Low,F.J.,1970,Astrophys.
        J.Lett.161,L101.
Gerola,H.,Iglesias,E.and Gamba,Z.,1973,Astron.Astrophys.
        24,369.
Harrington,J.P.,1968,Astrophys.J.152,943.
Harrington,J.P.,1969,ibid.156,903.
Harrington,J.P.and Marionni,P.A.,1976,Astrophys.J.
        206,458.
Hutchings,J.B.,1972,Monthly Notices Roy.Astr.Soc.
        158,177.
Kraft,R.P.,1964,Astrophys.J.139,457.
Lynden-Bell,D.and Pringle,J.E.,1974,Monthly Notices
        Roy.Astr.Soc.168,603.

MacAlpine,G.M.,1972,Astrophys.J.175,11.

Malakpur,I.,1973,Astron.Astrophys.24,125.

Mallik,D.C.V.,1975,Astrophys.J.197,355.

McKee,C.F.and Tarter,C.B.,1975,Astrophys.J.202,306.

Mustel,E.R.and Baranova,L.I.,1965,Astron.Zhurnal,42,42.

Mustel,E.R.and Boyarchuk,A.A.,1970,Astrophys.Space
          Sci.6,183.

Nariai,K.,1974,Publ.Astr.Soc.Japan 26,57.

Osterbrock,D.E.,1974,'Astrophysics of Gaseous Nebulae',
          San Francisco:Freeman and Co.

Paczynski,B.,1976,in'Structure and Evolution of Close
          Binary Systems',IAU Symp.73,D.Reidel,in press.

Payne-Gaposchkin,C.and Gaposchkin,S.,1942,Harvard
          Circular,No.445.

Pottasch,S.,1959a,Ann.Astrophys.22,310.

Pottasch,S.,1959b,ibid.,394.

Pottasch,S.,1959c,ibid.,412.

Schwarz,J.,1973,Astrophys.J.182,449.

Shakura,N.I.and Sunyaev,R.A.,1973,Astron.Astrophys.
          24,337.

Starrfield,S.,Sparks,W.M.and Truran,J.W.,1974a,Astro-
          phys.J.Suppl.28,247.

Starrfield,S.,Sparks,W.M.and Truran,J.W.,1974b,Astro-
          phys.J.192,647.

Tylenda,R.,1976,in'Multiple Periodic Variables',IAU
          Coll.29,Budapest,in press.

Tylenda,R.,1977,in preparation.

Williams,R.E.,1970,Astrophys.J.159,829.

ABUNDANCES DETERMINATIONS IN THE NEBULAR STAGE OF NOVAE.

S. Collin-Souffrin

Observatoire de Meudon, D.A.F., 92190 Meudon, France.

Introduction.

In 1959, Mustel and Boyarchuk analyzed the spectrum of
Nova Herculis 1934 at the maximum, by the method of the curve of
growth, and Pottasch analyzed in a serie of papers, the nebular
spectra of five novae. They came to the same conclusion : nitrogen and oxygen seem to be more abundant with respect to hydrogen
than in the atmospheres of "normal" stars.

Since Pottasch's study, very few other determinations of
abundances have been performed for the nebular stage of novae.
Indeed these computations require a very complete set of spectroscopical data and, in spite of a careful study of the physical
conditions in the envelope, lead to rather uncertain values.

The first part of this paper is devoted to a critical
analysis of the methods used to compute element abundances in
the nebular stage of novae, with an estimate of the "error-bars".
In the second part, we give the very scarce results obtained
until now.

The method and its limitation.

a) Generalities. The method has been used extensively for
planetary nebulae. The limitations are the same, but the following
specific problems do appear in addition.

1. The envelope is not spatially resolved, and this has many
important implications. First, the spectra of the star and of

M. Friedjung (ed.), Novae and Related Stars, 123-132. All Rights Reserved.
Copyright © 1977 by D. Reidel Publishing Company, Dordrecht, Holland.

the envelope are mixed. It is thus very difficult to separate
the continuous contributions of the 2 components, and to get the
Balmer discontinuity or the colour temperature of the emitting
nebular gas. Second, contrary to the case of planetary nebulae,
the geometrical dilution factor and the volumic filling factor
are not known directly from the observations.

2. Due to the high velocity of the envelope (sometimes
there are several velocity-systems), the emission lines are
blended and cannot be easily separated ; it is also difficult to
separate the continuum from the line emission spectrum. This
explains why the spectroscopical data differ sometimes strongly
from one observer to the other.

3. It is likely that the fluctuations of density are greater
in novae than in planetary nebulae, as is suggested from the
existence of different velocity systems. Also, the temperature
fluctuations are certainly more important, as I will discuss
later on.

The abundances computed in planetary nebulae by different
authors differ sometimes by a factor of ~ 1.5 for some elements
such as oxygen and neon and by a large factor for some others
like nitrogen and sulfur ; then the abundances obtained for
novae must be taken very cautiously and should be considered as
being essentially qualitative until now.

b) The method. The absolute intensity I_λ of a line λ measured
at earth is :

$$I_\lambda \ (\text{ergs cm}^{-2} \text{ s}^{-1}) = 1/4\pi d^2 \int f_\lambda (T_e, n_e) \ n(X_{i,j}) \ dV \qquad (1)$$

where λ is the wavelength of the line, d is the distance of the
nova, V is the volume of the envelope. $n(X_{i,j})$ is the numerical
density of the element X_i, in the j-state of ionization, giving
rise to the line λ, $f_\lambda(T_e, n_e)$ is the emissivity of the line,
depending on the physical conditions (T_e, electronic temperature,
n_e, electronic density) and on the atomic parameters.

The intensity must be corrected for the reddening which, in
some cases, seems more important than it would be, due to the
contribution of the interstellar medium alone. Notice also that
this formula accounts only for optically thin lines ; the Balmer
lines, in the first nebular stages, should also be corrected for
self absorption effects (cf. Pottasch, 1959).

The function f is strongly dependant on T_e for the forbidden
lines. For the permitted recombination lines and for the forbidden
lines at low density ($n_e \lesssim 10^4$) it is proportional to n_e and for

the forbidden lines at high density ($n_e \gtrsim 10^8$) it is a constant (the density in the nebular phase of novae lies between these values).

The difficulties to get accurate ionic abundances appear immediately in equation 1. Equation 1 is approximated by :

$$N(X_{i,j})/N(H) = \left[X_{i,j} \right] = (I_\lambda / I_{H\beta}) \, (f_{H\beta}/f_\lambda) \qquad (2)$$

where $\left[X_{i,j} \right]$ is the ratio of the number of ions $X_{i,j}$, to the total number of hydrogen nuclei.

Equation 2 has several implications :

1. The lines are emitted only by the ionized part of the envelope. This is probably the case for Hβ and for the lines due to ionized species (except perhaps for species having low ionization potentials, such as S^+), but it is not necessarily so for the lines due to neutral species : indeed these lines can be emitted in a collisionally heated neutral gas.

2. The density is uniform : actually this implication is certainly not verified, since the various methods used to determine the density give results differing by a factor 2 or 3 (for instance, cf. Malakpur, 1973, for Nova Delphini 1967).

Moreover the density seems to vary from one velocity system to the other (cf., always for Nova Delphini 1967, Gallagher and Anderson, 1976).

3. The implied temperature is uniform. This again is probably a bad approximation, since models computed for planetary nebulae have shown that it is not the case. The trouble is then that the forbidden lines come mainly from the hottest part of the regions containing their emissive ions ; they may as well originate in different regions, with very different temperatures. This effect should be emphasized in novae, since the most important cooling agents, i.e. N and O, are found to be highly overabundant. For instance, the temperature of the region emitting the [OII] and [NII] lines is certainly different from that emitting the [OIII] and [NeIII] lines, if the two regions are separated, like in planetary nebulae and in HII regions (cf. Collin-Souffrin and Joly, 1976).

But the most crucial approximation occurs when one determines the total element abundances, $\left[X_i \right] = \Sigma_j N(X_{i,j})/N(H)$. In this formula, the abundances of many important ions such as O^{+3} and N^{+3} cannot be deduced from the observations, and their amount must be estimated theoretically.

Until now, two methods have been suggested for this estimation.

Ruusalepp and Luud (1970) proposed to use the modified Saha equation. This should be a very rough approximation. In fact, it is well known that this equation does not take into account how the absorption coefficients of different elements depend upon ν. On the other hand, owing to the large overabundance of CNO in the photosphere of the nova, the ionizing spectrum should differ drastically both from a blackbody spectrum and from the spectrum of the nucleus of a planetary.

Pottasch follows a more empirical scheme, which certainly leads to better values of the abundances. He argues that, during the whole period of observation, the population of one stage of ionization of an element will reach a maximum value, which can be considered as the total abundance of the element ; this gives a lower limit on the amount of the element. Actually Pottasch claims that, for oxygen, all the stages of ionization considered separately give the same result (within 25 %). However, this method requires to take very carefully into account the optical thickness of the Balmer lines at the beginning of the nebular phase, as he has himself done.

Finally, I should mention the different methods used to determine the very important parameters T_e and n_e.

1. For Nova Delphini 1967, Ruusalepp and Luud, and for Nova Herculis 1963 Doroshenko (1968) essentially used the continuum emission (assumed to be a superposition of a blackbody spectrum with an optically thin gas spectrum), and the ratio of the [OIII] lines $I(4363)/I(4959 + 5007)$.

2. Pottasch (1959, 1967) used the following method :

- he uses the velocity as a function of time to deduce R (the radius of the envelope, assumed spherical), ΔR (its thickness, if its expands with a velocity equal to three times the sound velocity), and V (its volume).

- he shows that the whole volume of the envelope is ionized (except perhaps at the very beginning of the nebular phase), and then he is able to deduce n_{H^+} from the intensities of the Balmer recombination lines. The total density n_H and the electronic density n_e can be computed ; the contribution of He^+ and He^{++} is deduced from the recombination lines of HeI and HeII.

This method enables to set a lower limit to the density, since it assumes that the filling factor of the shell is equal to unity. Malakpur (1973) has shown that a direct determination

of n_e, by using the ratios of auroral to nebular lines of different ions gives for Nova Delphini 1967 a value of n_e only two
times higher (in 1968 and 1969). Notice that this last method
can also be criticized, since it assumes that all the lines
(auroral and nebular, due to O^{++}, O^+, N^+...) are emitted in an
homogeneous region.

The knowledge of the density leads to T_e, by using the ratio
[OIII] 4363 / (5007 + 4959). Since n_e is underestimated, the
temperature can be <u>overestimated</u>. This is probably one way of
explaining the very high temperature found in the post-novae
stage (cf. Andrillat and Collin-Souffrin, a communication in
this conference), if the envelope is highly inhomogeneous during
this last phase.

As a conclusion to this discussion, I think that the <u>only</u>
<u>appropriate method to determine element abundances in the envelope</u>
<u>of novae would be to build a self-consistent model</u>. It should
include the computation of the stage of ionization of the envelope,
its evolution with time, and the study of the thermal equilibrium
of the gas (by using the abundances as parameters), and finally
the comparison of the computed line intensities with the observed
ones. This is a very difficult problem, since the star temperature
and luminosity as functions of time are not known, and were they
known, the ionizing spectrum in the far UV would be hard to find
(cf. Balick and Sneden, 1976, who have shown how important the
CNO abundances are for the UV spectrum of a hot star). Moreover,
this is not a quasi stationary problem since the characteristic
times of evolution (cooling, recombination) are larger than the
duration of the nebular stage, at least in the late phases. The
spherical symmetry and the homogeneity of the envelope are also
far from being realised. And finally it is likely that the
emissive nebular spectrum is influenced by collisions (cf. Andrillat and Collin-Souffrin, this conference).

However, in spite of the difficulties, a self-consistent
model has been proposed by Tylenda (1976) for Nova Delphini
1967. I hope that others will follow.

Let us now examine the elements and try to estimate the
uncertainties on their abundances.

1. <u>Helium</u> : [He^+] and [He^{++}] can be determined from recombination lines. He^o is certainly unimportant, except at the beginning of the nebular phase, [He] is then deduced within the range
of uncertainty due to the line intensities (I would say 50 %
to 100 %).

2. Nitrogen : the only observed ions are N^+ (forbidden lines) and N^{++} (the $\lambda 4640$ line, whose mechanism of emission is not well-known - recombination or fluorescence). The N^+ ion has a small relative abundance, except at the beginning of the nebular phase - and in this phase, the Balmer lines are probably optically thick, and their intensity should be computed very carefully.

Some authors (Doroshenko, 1968 ; Ruusalepp and Luud, 1970) deduce the value of $[N^{+3}]$ from the intensity of the line NIII 4640 : this is based on the assumption that it is a recombination line, and it leads to an overestimation of the abundance.

Besides, computed models of planetary nebulae, HII regions, and non thermally ionized regions, have shown that the fractional amounts of N^+ and of O^+ are of the same order of magnitude.

3. Oxygen : generally O°, O^+ and O^{+2} are observed (forbidden lines). O^{+2} is the dominant ion at the beginning of the nebular phase, but O^{+3} becomes generally important after a few months. Its abundance can be computed if some fluorescence or recombination OIII lines are observed.

4. Other elements :

- Neon : Ne^{+2} and sometimes Ne^{+4} can be deduced from the forbidden line intensities, and in the second case, the abundance of neon is probably the best known one. The ratio $[Ne^{+4}]/[Ne^{+2}]$ can give some information on the ratio $[O^{+3}]/[O^{+2}]$.

- Carbon : the λ 4267 line of CII enables to compute the abundance of C^{+2}, with some assumed effective recombination coefficient, but this ion is never dominant.

- Sulfur : S^+ is observed, but this ion is also unimportant, except perhaps at the very beginning of the nebular phase : the same remark as for N^+ holds.

- Iron : $[Fe^+]$ can be deduced, but is not dominant. $[Fe^{+6}]$ can be well estimated, and, from the equation $[Fe^{+6}]/[Fe]$ $\sim [Ne^{+4}]/[Ne]$ one can deduce $[Fe]$ within some uncertainty (~ 3 ?). Finally the abundance of Fe^{+9} can also be computed if the coronal line $[FeX]$ 6374 is observed, but it probably refers to a very thin and hot region, and cannot be directly compared to Hα (cf. Andrillat and Collin-Souffrin, this conference).

THE RESULTS.

Table 1 summarises the results available until now.

Nova	[He]	[N]	[O]	[Ne]	[S]	[Fe]	Ref.
Per 1901	.43	2.2 -2	3.2 -4	2.1 -4	5.1 -7		2
	.17		4.1 -3		1.2 -3		1
Aql 1918	.53		2.8 -4				2
	.32		2.2 -2	1.4 -4			1
Pic 1925	.27	7.9 -3	1.7 -4	2.5 -4	8.5 -7		2
	.32		1.5 -3		1.2 -3		1
Her 1934	.51	8.3 -3	1.8 -3	4.3 -5			2
	.07	3.7 -3	3.5 -3	1.1 -4			1
Lac 1936	.24	2.8 -2	1.0 -4				2
	.06	7.8 -3	6.3 -3				1
Oph 1958	.43	2.5 -3	6.5 -3		1.0 -4	1.6 -4	3
	.31	1.4 -2	1.2 -6		1.5 -7		2
Lac 1950	.33	3.0 -3	1.6 -4		4.0 -7		2
	.25	3.0 -3	5.0 -3				4
Del 1967	.50	3.1 -3	2.4 -4				2
		5.0 -3	6.8 -3				5
	.25	> 1. -3	5.0 -3	1.0 -4		6.0 -5	4
Her 1963	.15	1.0 -2	5.0 -3	1.0 -4	> 4. -5	3.0 -5	6
	.20	4.5 -2	3.0 -3				7
	.19	3.0 -2	2.8 -3				8
	.27	5.3 -2					9

Table 1

Notes to table 1 :

References :

1. Pottasch, 1959.
2. Ruusalepp and Luud, 1970.
3. Pottasch, 1967.
4. Collin-Souffrin, 1976.
5. Tylenda, 1975.
6. Andrillat and Collin-Souffrin, 1976.
7. Barchak and Boyarchuk, 1965.
8. Doroshenko, 1968.
9. Kuningas and Luud, 1968.

(For Ruusalepp and Luud, $\left[O^{+2}\right]$, $\left[N^{+3}\right]$, $\left[S^{+}\right]$ and $\left[Ne^{+2}\right]$ are given instead of $[O]$, $[N]$, $[S]$ and $[Ne]$).

The names of the novae are in column 1. The 6 following columns show the computed abundances. The last column gives the references.

Nova Persae 1901, Nova Aquilae 1918, Nova Pictoris 1925, Nova Herculis 1934, Nova Lacertae 1936 and Nova Ophiuchi 1958, have been studied by Pottasch and by Ruusalepp and Luud. The discrepancies between their results are due to the fact that Pottasch did use the method previously described, to determine the total abundances, while Ruusalepp and Luud had only determined the abundances of some ions. Furthermore, their $\left[N^{+3}\right]$ is certainly obtained by the method of the recombination line, and therefore is overestimated. The large discrepancy on $[S]$, which is deduced from $\left[S^{+}\right]$ in both cases, probably arises from the fact that Ruusalepp and Luud do not use the lines $[SII]$ at the very beginning of the nebular phase, contrary to Pottasch, who takes into account the optical thickness of the Balmer lines at this moment. Then, one can consider that his results are much closer to the real values.

For Nova Lacertae 1950, my results have been obtained by a method similar to that of Pottasch, and by using the very extensive data published by Larsson-Leander (1954). However, the result on $\left[S^{+}\right]$ does not correspond to the very beginning of the nebular phase, and is not reliable.

Tylenda (1975) has computed the abundances of Nova Delphini 1967 with a self consistent model of thermal and ionization equilibrium of the envelope (his paper in this conference). I have also studied this nova using the data of Andrillat, Fehrenbach and Houziaux (1974) : all these results will be published in greater details elsewhere (Collin-Souffrin, 1976).

Finally, for Nova Herculis 1963, Andrillat and Collin-Souffrin (1976) and Doroshenko (1968) have obtained similar results.

These results show some general features. The abundance of helium, although not accurately determined (cf. the discrepancies between different authors) seems to be systematically high, with a mean value $[He] = 0.25$. If one considers the best reliable data only, oxygen seems about 10 times overabundant ($[O] = 5\ 10^{-3}$), and nitrogen about 30 times overabundant ($[N] \sim 3\ 10^{-3}$). Neon does not seem abnormal, while sulfur (only Pottasch's results are available) seems highly overabundant. Iron is perhaps slightly overabundant.

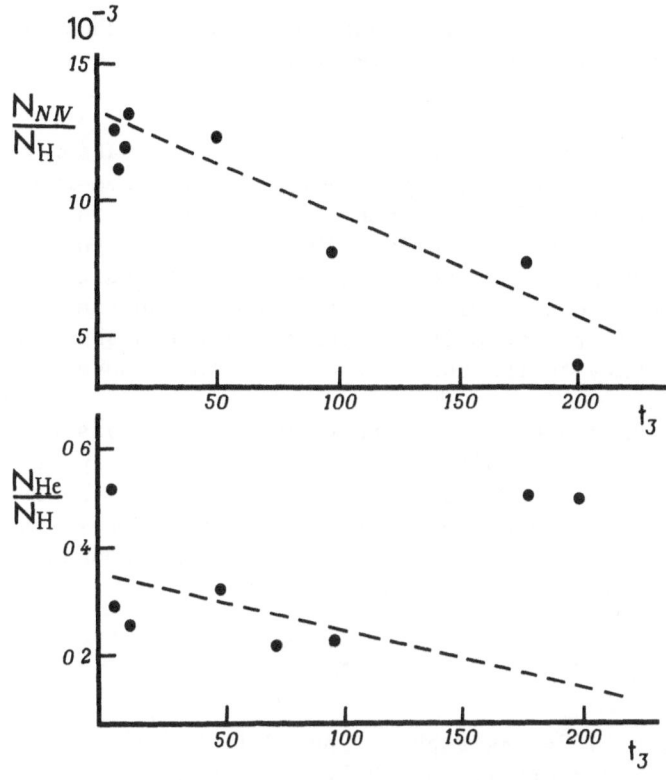

Figure 1

Ruusalepp and Luud have shown the existence of a correlation between $[N^{+3}]$ and t_3 (the time for a decrease of 3 magnitudes), and between $[He]$ and t_3. This appears on figure 1.

However, there may be no real correlation between the abundances and the time scale of decrease. Indeed, a difference in the stage of ionization of the shell (due to a difference in the temperature of the star, or in the density of the envelope, or to any other cause) for slow and rapid novae, may account for the $[N^{+3}]/t_3$ correlation : it is then an interesting result, even if it has nothing to do with abundances. The second relation $[He]/t_3$ is not well established since He - abundances are widely dispersed.

Finally, I should mention a preliminary result that I have obtained by studying the abundances in the "post novae stage" : the abundances seem to decrease with time. As an example, in Nova Herculis 1934, the abundance of oxygen deduced from the data published by Swing and Jose (1949) seems to decrease from $3.5 \ 10^{-3}$ in 1934 to $1.6 \ 10^{-3}$ in 1940, $8.5 \ 10^{-4}$ in 1942 and $2.5 \ 10^{-4}$ in 1947 and 1949. A similar tendency appears for Nova Delphini 1967, from 1970 to 1975, for RR Telescopii from 1961 to

1972 (spectroscopical data of Aller and Kayes, 1974), and for
Nova Persae 1901 (spectroscopical data compiled by Mc Laughlin,
1949).

Is this an artefact due to the method of determination of
the abundances ? In this method, the temperature deduced from the
[OIII] lines, assuming a uniform and decreasing density, increases
strongly with time (it reaches several 10^4 K), leading to a
decrease of the computed abundances. It is very difficult to
explain such a decrease by a mixing of the shell with the interstellar medium, since the velocity of the shell would then be
much reduced, contrary to the observations for these novae.

The alternative explanation is that the density does not
remain uniform, but becomes highly inhomogeneous in the post
nova stages (with a local density reaching $\sim$ 100 times the mean
value). This would lead to a low temperature deduced from the
[OIII] lines, and therefore to the constancy of the abundances.
Further studies should be developped on this problem.

Bibliography.

Aller, L., Kayes, C.,1974, Astrophys. Spa. Sci. 30, 387.
Andrillat, Y., Collin-Souffrin, S. 1976, in preparation.
Andrillat, Y., Fehrenbach, C., Houziaux, L. 1974, Astrophys. Spa.
      Sci. 31, 169.
Balick, B., Sneden, C. 1976, Astrophys.J. 208, 336.
Bartasch, T.M., Boyarchuk, A.A. 1965, Isv. Kremskoï Obs. 33, 173.
Collin-Souffrin, S. 1976, in preparation.
Collin-Souffrin, S., Joly, M. 1976, Astron. Astrophys., in press.
Doroshenko, V.T. 1968, Astron. Zh. 45, 121.
Gallagher, J.S., Anderson, C.M. 1976, Astrophys.J. 203, 625.
Kuningas, S., Luud, L. 1968, Tartu Obs. Publ. 36, 222.
Larsson-Leander, G. 1954, Stockholm Obs. Ann. 18, 4.
Mc Laughlin, D.B. 1949, Publ. Obs. Michigan, 9, 13.
Malakpur, I. 1973, Astron. Astrophys. 24, 125 and 28, 393.
Mustel, E.R., Boyarchuk, M.E. 1959, Astron. Zh. 36, 762.
Pottasch, S.R. 1959, Ann. d'Astrophys. 22, 297, 394 and 412.
Pottasch, S.R. 1967, Bull. Astron. Inst. Neth. 19, 227.
Ruusalepp, M., Luud, L. 1970, Tartu Obs. Publ. 39, 89.
Swing, P., Jose, P.D. 1949, Astrophys.J. 110, 475.
Tylenda, R. 1975, European Conference on Astronomy, Leicester,
      England.
Tylenda, R. 1976, Thesis.

SPECTRAL EVOLUTION OF NOVA HR DEL DURING ITS DECLINE TOWARDS MINIMUM.-

P. RAFANELLI , L. ROSINO.-

ASIAGO ASTROPHYSICAL OBSERVATORY OF THE UNIVERSITY OF PADUA.-

The purpose of the present report is to summarize the spectral evolution of Nova HR Del during its declining stage;namely in the period May 1968,July 1975.The evolution of the spectrum was as follows:-
1968:-The excitation was increasing after the last maximum,occurred on May.The low excitation emission lines of FeII,SiII,TiII,CaII, bordered on their blue side by strong PCyg absorptions faded away. To the contrary the emission features of HeI,NII,OII and the forbidden lines of [NII]5754,[OI]6300-64 strengthened.The nova entered the nebular stage near end July,3.5 mag below maximum.Noticeable was the presence at this early phase of many forbidden lines of [FeVI]and [FeVII].The highest excitation line was [FeX]6374,blended with [OI] 6364.The profiles of $H_\beta,H_\gamma,H_\delta$ and [OIII]4363 showed three components in September.Their radial velocities were:-415+12 Km/s;-16+13 Km/s; +461+35 Km/s.At the same time H_8 and [NeIII]3869 were still more complex with six components;their velocities were:-458+16 Km/s;-298+6 Km/s;-92+9 Km/s;+99+20 Km/s;+314+6 Km/s;+485+1 Km/s.
1969:-The excitation was still increasing.The HeII lines were becoming stronger and HeI weaker.$H_\beta,H_\gamma,H_\delta$ and [OIII]4363 showed in June four peaks,having the radial velocities:-447+29 Km/s;-101+15 Km/s; +130+21 Km/s;+457+49 Km/s.H_8 and [NeIII]3869 were alway present with six components,having velocities:-457+13 Km/s;-297+13 Km/s;-104+4 Km/s;+113+8 Km/s;+237+10 Km/s;+513+15 Km/s.
1970-71:-[FeX]6374 was no more present.Many forbidden and high excitation lines split into two components,having radial velocities: -436+38 Km/s;+381+40 Km/s.
1972-73:-All the lines faded relatively to the continuum.HeI,[FeVI], [FeVII]and [NII]5754 disappeared.The radial velocities of the peaks of the double lines were:-410+94 Km/s;+383+58 Km/s.
1974-75:-No important changes were observed in the spectrum at minimum.The strongest emission lines were:H_α,[OIII]5007-4959,HeII 4686.-

M. Friedjung (ed.), Novae and Related Stars, 133. All Rights Reserved.
Copyright © 1977 by D. Reidel Publishing Company, Dordrecht, Holland.

THE LIGHT CURVE OF DQ HERCULIS 1934

P.B. Bosma

Department of Physics and Astronomy,
Free University, Amsterdam

The light curve of DQ Herculis 1934 shows after a primary maximum with an apparent brightness of 2^m0, a slow decrease of 2^m5 in some 100 days, then suddenly it drops 8^m6 in 30 days to a deep minimum. After this minimum it rises for about 120 days until it reaches a secondary maximum of about 6^m4 above the minimum. From this time on a slow final decline sets in.

A first attempt to explain the rise of the light curve from the minimum to the secondary maximum has been made by Grotrian in 1937. His model consists of a static hydrogen shell ejected by the active star. An assumed sudden increase in the Lyman continuum brightness of the central star at the time of the minimum in the light curve, gives rise to a temporary increase in the degree of ionization of the shell, until after some time a new state of ionization equilibrium is reached. Recombinations in the shell cause nearly all Lyman continuum photons to be converted to photons of lower energies, from which a large part will be observable as visible light. As a result, an increase in the brightness of the shell occurs.

The idea of a sudden increase in the brightness of the central star seems a bit hypothetical and is avoided in the model described here. On the other hand, use will be made of the fact that the shell is expanding. It is assumed that from the moment the steep decrease to the minimum sets in, the central star has come to rest and that from that time on the number of photons, emitted by the star in the Lyman continuum, remains constant. Before that time the high luminosity of the nova can possibly be explained as caused by a continuing eruptive activity of the star, sustaining a high degree of ionization throughout the shell. Thereafter the ioniza-

M. Friedjung (ed.), Novae and Related Stars, 135-136. All Rights Reserved.
Copyright © 1977 by D. Reidel Publishing Company, Dordrecht, Holland.

tion conditions in the shell will be governed only by the Lyman
continuum radiation of the star, and if the optical thickness of
the shell is large enough, the ionization will be restricted to the
Strømgren region. As a consequence, the outer layers will have to
recombine. The calculations presented by Bosma(1975a) show that in
the outer parts of the shell a considerable number of H⁻ions and
hydrogen atoms with electrons in their third and higher energy
levels, originating from recombinations and radiative and collisional transitions, can give rise to an appreciable absorption
coefficient for visible light. So, apart from the fact that the
visible radiation, coming from recombinations in the ionized part
of the shell, decreases because of the shrinking of this ionized
region, the radiation can further be diminished by absorption in
the outer layers, thus giving rise to a deep minimum in the light
curve. The expansion of the shell however, causes the particle density to decrease, which in turn gives rise to a shift of the ionization front towards the outer edge of the shell, the absorption in
the outer layers decreases and whereas the output of visible light
from the Strømgren region is constant, as long as this region lies
within the shell, the brightness of the nova starts to rise.

In the calculations performed to construct the light curve of
DQ Her 1934, the degree of ionization in an expanding, spherical,
uniform hydrogen shell surrounding the central star has been calculated as a function of position in the shell and time. The equation used is of the type:

$$\frac{df}{dt} = A(1-f) - Bf^2 + C(f - f^2),$$

where f is the degree of ionization. The first term on the right
hand side stands for the influence of the star radiation, the
second term describes the effect of recombinations in the shell and
the last term gives the ionizations caused by electron-atom collisions. The first two terms are mainly of interest in the ionized
shell region, the last term dominates in the outer weakly ionized
zones. Details of the calculations are given by Bosma(1975a).

A choice of 10^{14} cm for the outer shell radius, 10^{10} cm^{-3} for
the particle density, an electron temperature of 10^{13}K and, for the
central star, a Lyman continuum photon emission of 6×10^{47}s^{-1} as
starting values for the integration of the ionization equation,
leads to a very good agreement between the calculated and the observed light curve.

REFERENCES

Bosma, P.B. 1972, Astron & Astrophys., vol.21, p.223.
Bosma, P.B. 1975a, Astron & Astrophys., vol.40, p.175.
Bosma, P.B. 1975b, Astron & Astrophys., vol.40, p.185.

EVIDENCES FOR THE DEVELOPMENT OF A COLLISIONAL SPECTRUM IN THE NEBULAR STAGE OF NOVAE.

Y. Andrillat and S. Collin-Souffrin.

Observatoire de Haute Provence, 04300 Forcalquier, France.
and
Observatoire de Meudon, D.A.F., 92190 Meudon, France.

Coronal lines in the novae spectrum.

The study of the coronal lines of $[FeX]$ and $[FeXIV]$ in the spectrum of Nova Ophiuchi 1958, Nova Herculis 1963, Nova Delphini 1967, shows that a very thin and hot ($\gtrsim 10^6$ K) region develops in the envelope.

Following an analysis described in a previous paper, in this conference, we have studied the evolution of the ionic abundances in Nova Herculis 1963. The results are summarized in Table 1.

t	$[\bar{n}_e]$	$[T_e]$	$[He^+]$	$[He^{++}]$	$[O^0]$	$[O^+]$	$[O^{2+}]$
2	10^8	$< 10^4$	.06	.005	$<3\ 10^{-4}$	$< 10^{-2}$	$< 10^{-2}$
3	$4\ 10^7$	10^4	.10	.022	$3\ 10^{-5}$	$2.4\ 10^{-4}$	$2.2\ 10^{-3}$
4	$2\ 10^7$	10^4	.15	.040	$3\ 10^{-5}$	$2.1\ 10^{-4}$	$2.5\ 10^{-3}$
5	10^7	$1.35\ 10^4$	.10	.040	$2.2\ 10^{-5}$	$2\ 10^{-5}$	$1.5\ 10^{-3}$
6	$4\ 10^6$	1.4 to $1.9\ 10^4$	.13	.040	$6\ 10^{-6}$	10^{-5}	$6\ 10^{-4}$

Table 1

t : is given in months ; $[X_i] = n(X_i)/n(H)$.

It can be shown that the ionic abundances (and even those of Fe^{+5} and Fe^{+6}) are typical of a radiative ionization, while the coronal spectrum which develops during the 5th month is due to a collisional ionization. The volume of the collisional region is at most 10 % of the total volume and its temperature is of the order 10^6 K (if the hot region is compressed, the volume is even

M. Friedjung (ed.), Novae and Related Stars, 137-138. All Rights Reserved.
Copyright © 1977 by D. Reidel Publishing Company, Dordrecht, Holland.

smaller).

A similar conclusion is reached for other novae showing coronal lines.

For Nova Ophiuchi 1958, by using the intensities of the coronal lines given by Pottasch (1967) one finds that the number of "hot" ions increases with time (cf. Table 2).

t(months)	$N(Fe^{+9} + Fe^{+13})/N(H^+)$ case a : $T_e = 1.6 \ 10^6$ for $[FeX]$ and $[FeXIV]$	$N(Fe^{+9} + Fe^{+13})/N(H^+)$ case b = $T_e = T_{max}^{(1)}$ for $[FeX]$ and $[FeXIV]$
1.3	$.77 \ 10^{-5}$	$1.54 \ 10^{-5}$
2	$.90 \ 10^{-5}$	$3.2 \ \ 10^{-5}$
3	$1.4 \ \ 10^{-5}$	$3.85 \ 10^{-5}$

(1) T_{max} = temperature of the maximum emissivity of the line (cf. Nussbaumer and Osterbrock, 1970).

Table 2

Post nova stage.

The study of the post nova stage of Nova Persei 1901, Nova Herculis 1934 and RR Telescopii, based on the spectroscopical data published respectively by Mc Lauhglin (1949), Swing and Jose (1949), and Aller and Kayes (1974), shows that, after some years the temperature deduced from the $[OIII]$ lines reaches several 10^4 K, while at the same time, the stage of ionization is rather decreasing.

This effect can be interpreted as a decrease of the radiative ionization and an increase of the influence of collisions. But it also leads to the conclusion that the element abundances decrease with time (cf. the paper by Collin–Souffrin, in this conference) a fact very difficult to explain.

An alternative interpretation would be that the method of determination of the temperature, based on the assumption of a uniform density in the shell, is completely erroneous, and that a strong clumpiness develops in the shell (with a filling factor $\sim 1\%$).

Bibliography.
Aller, L., Kayes, C. 1974, Astrophys. Spa. Sci. 30, 387.
Mc Laughlin, D.B. 1949, Publ. Obs. Michigan, 9, 13.
Nussbaumer, H., Osterbrock, D.E. 1970, Astrophys.J. 161, 81.
Pottasch, S.R. 1967, Bull. Astron. Inst. Neth. 19, 227.
Swing, P., Jose, P.D. 1949, Astrophys.J. 110, 475.

THE SLOW NOVA RR TEL AND THE SYMBIOTIC STARS

M.W. Feast

S.A.A.O. Cape.

The slow nova RR Tel belongs to a group of symbiotic stars which contain in addition to an M type component, a hot dust shell. For symbiotics without dust, infrared photometry shows that the M component is constant in light. For these systems the mass exchange is usually considered to be through the inner Lagrangian point. For the dusty systems it is found that the cool stellar component shows large variations in light. This leads to their identification as Mira variables. For a system such as RR Tel therefore the companion may well be accreting from a general stellar wind from the Mira. RR Tel and similar systems show much stronger radiation from dust than normal Miras. Either an unusually dense stellar wind is needed to produce a system of this kind or such a system produces dust, perhaps in a high density region resulting from the inter-action of the stellar wind with the companion.

Details of this work are given in a paper entitled "The Infrared Variability and Nature of Symbiotic Stars" by M.W. Feast, B.S.C. Robertson and R.M. Catchpole (M.N.R.A.S. in press).

PART III

<u>(b) TRANSIENT X RAY SOURCES</u>

OPTICAL PROPERTIES OF X-RAY NOVAE

F.Ciatti , A.Mammano , A.Vittone

Asiago Astrophysical Observatory

Several transient X-ray phenomena have been observed, whose X-ray light curves closely resemble those of classical novae at optical wavelengths. We report spectroscopic observations of the stars V 616 Mon and HD 245770 (BD+26°883), suggested optical counterparts of the transient sources A 06.20-00 and A 05.35+26. Our spectrograms have been obtained with intermediate dispersion in the wavelength range 3800-10900 Å.

In V 616 Mon we observe a decrease of excitation, pertinent to the O class in Sept.1975, and to an intermediate B star in Feb. 1976. Cooler features are also recorded. The star, sometimes called Nova Mon 1975, can be excluded from novae according to its spectral evolution, and properties of the radio and X-ray source. We present several evidences of a binary nature with a late K and likely a neutron star. In this low mass system the X-ray outburst has been followed by optical reverberation.

HD 245770 can be classified as a giant early B star with emission lines. Variations of Balmer emission and HeI absorption lines have been observed (Nov.1975-Apr.1976). Together with irregular light variations, they support the identification. Also in this case a binary system is suggested. During an X-ray outburst similar to that in A 06.20-00, this star was constant because of its higher temperature and luminosity.

The division of X-ray novae in two classes, presenting different X-ray and optical properties, is discussed with their possible evolutionary phase and galactic distribution.

The full paper will be submitted to Astronomy & Astrophysics.

M. Friedjung (ed.), Novae and Related Stars, 143-144. All Rights Reserved.
Copyright © 1977 by D. Reidel Publishing Company, Dordrecht, Holland.

Duerbeck,H.W. and Walter,K.(1976) Astr.Ap. 48, 141.
Matilsky,T., Bradt,H.V., Buff,J., Clark,G.W., Jernigan,J.G., Joss,
    P.C., Laufer,B., McClintock,J. (1976) Preprint.
Robertson,B.S.C., Warren,P.R., and Bywater,R.A. (1976) Inf. Bull.
    Var.Stars No. 1173.
Wu,C.C., Aalders,J.W.G., van Duinen,R.J., Kester,D., Wesselius,P.R.
    (1976) Astr.Ap. 50, 445.

THE 8-DAY MODULATION IN V616 MON (A0620-00)

C.Chevalier[1], E.Janot-Pacheco[1], H.Mauder[2] and S.A.Ilovaisky[3]

(1) Observatoire de Paris, Meudon, France
(2) Astronomical Institute, Tübingen University, FRG
(3) Centre d'Etudes Nucléaires de Saclay, France

Photoelectric photometry of V616 Mon, the optical counterpart of the bright transient X-ray source A0620-00, obtained at the Haute Provence Observatory in October 1975 and at the European Southern Observatory in March 1976 shows evidence of a low-amplitude 8-day intensity modulation. Walter and Duerbeck (1976) have reported a 4-day periodicity soon after outburst. Interpreting their results in terms of a modulation having twice that period, and thus showing two maxima and two minima per cycle, and combining them with our data, we obtain a period of 7.80 ± 0.03 days. This modulation agrees in period and probably in phase with that detected in X-rays in January 1976 (Matilsky et al. 1976). Its amplitude increases with wavelength in March but no such effect is detectable in October. We note a similar lack of correlated color variability in the Duerbeck and Walter data. A recurring eclipse-like feature, which may be phase-locked to the 7.8-day cycle (Robertson et al. 1976), is superposed onto one cycle of the modulation in March. Significant intra-night variability is usually present, amounting to as much as 0.17 mag in two hours. Such variability, also detected at u.v. wavelengths (Wu et al. 1976) may be due to an underlying 1-day periodicity. Possible interpretations of both the 7.8-day period and the short time-scale variability have been examined. Our tentative conclusion is that the 7.8-day modulation is probably due to orbital motion. X-ray heating seems to be ruled out as a probable mechanism for the modulation in view of the inverse color effect observed (bluer when fainter). Perhaps the most direct evidence in favor of the binary nature of the A0620-00/V616 Mon system comes from the existence of phase-locked eclipse-like features.

M. Friedjung (ed.), Novae and Related Stars, 145. All Rights Reserved.
Copyright © 1977 by D. Reidel Publishing Company, Dordrecht, Holland.

THE X-RAY AND OPTICAL LIGHT CURVES OF A0620-00 (V616 MON)

K.Pounds[1], M.Watson[1], C.Chevalier[2] and S.A.Ilovaisky[3]

(1) Physics Department, Leicester University, UK
(2) Observatoire de Paris-Meudon, France
(3) Centre d'Etudes Nucléaires de Saclay, France

A complete X-ray light curve for the transient source A0620-00 has been constructed using Ariel V data. This curve is characterized by a sharp initial rise with a pre-maximum "dip" (Elvis et al. 1975) and by a two-sloped flux decrease (0.034 mag day^{-1} up to October 1975 and 0.053 mag day^{-1} thereafter). In early 1976 a factor of ten increase in flux lasting for about two months dominates the light curve until the final, abrupt decline around mid-March. Two other small flux increases were detected, a short one in September 1975 (20% increase) and a longer one in October 1975 (50% increase) just at the time of the change in slope. Several intensity minima are present in the light curve near minimum flux and are probably related to the 8-day cycle reported by Matilsky et al. (1976) and Chevalier et al.(1976). The amplitude of these minima seems to increase with decreasing flux level. An exception is a sharp "dip" (25% decrease) observed in November 1975.

A composite optical light curve for V616 Mon has been constructed from all available post-outburst observations, including photoelectric, photographic and visual observations. This curve is characterized by a much smaller rate of decrease (0.015 mag day^{-1} averaged over 1975) and by a secondary maximum in early 1976 which coincides perfectly with that seen in X-rays, including the final, abrupt decline. There is marginal evidence for an optical flux increase at the time of the October X-ray "hump".

Periodic low-amplitude optical modulation in the light curve has been reported by Duerbeck and Walter (1976) and Chevalier et al.(1976). At least two eclipse-like features have been observed (Robertson et al. 1976, Chevalier et al.) which may be phase-locked to the low-amplitude modulation. A detailed search for other such features is made difficult by the inhomogeneous time coverage and unequal quality of the optical data.

Chevalier,C., Janot-Pacheco,E., Mauder,H., and Ilovaisky,S.A.
    (1976) Paper presented at this meeting.
Duerbeck,H.W., and Walter,K. (1976) Astr.Ap. $\underline{48}$, 141.
Elvis,M., Page,C.G., Pounds,K.A., Ricketts,M.J., and Turner,M.J.L.
    (1975) Nature $\underline{257}$, 656.
Matilsky,T., Bradt,H.V., Buff,J., Clark,G.W., Jernigan,J.G., Joss,
    P.C., Laufer,B., and McClintock,J. (1976) Preprint.
Robertson,B.S.C., Warren,P.R., and Bywater,R.A. (1976) Inf. Bull.
    Var. Stars No. 1173.

SPECTROSCOPIC OBSERVATIONS OF V616 MON (A0620-00)

S.A.Ilovaisky[1] and C.Chevalier[2]

(1) Centre d'Etudes Nucléaires de Saclay, France
(2) Observatoire de Paris, Meudon, France

We report on spectroscopic observations made in March 1976 at the time of the optical and X-ray secondary maximum of V616 Mon (A0620-00). The three highest signal-to-noise-ratio spectra, having a resolution of 3 Å and covering the spectral interval $\lambda\lambda$ 3900-5600 Å were obtained near phase ϕ = 0.5 (maximum flux) in the 8-day cycle (Chevalier, Janot-Pacheco, Mauder and Ilovaisky 1976) when the object was near B = 13.65, B-V = 0.32, U-B = -0.50. The spectra were obtained at the coudé focus of the 1.52-m European Southern Observatory reflector using the Echelec spectrograph in single-spectrum mode and a 30-mm Lallemand electronographic camera. After digital processing with a PDS microphotometer, the three best spectra were numerically added. The composite spectrum shows the following weak (10% of the continuum), broad (50 Å) features: (a) Balmer lines (H_β, H_γ, H_δ, H_ϵ) in absorption, with narrow, weak emission cores; (b) He II λ4686 Å in emission with structure in the blueward wing, probably due to weaker CIII-NIII $\lambda\lambda$ 4630-4650 Å emission; (c) He I absorption lines at λ 5015 Å and λ4471 Å, the latter certainly blended with the diffuse interstellar absorption band near λ 4430 Å. Radial velocity measurements, though somewhat uncertain due to the weakness and shallowness of the lines, indicate no evidence whatever for an expansion velocity greater than about 150 km s^{-1}. We note satisfactory agreement with spectra taken in January 1976 with the 3.9-m AAT reflector and a Wampler-Robinson IDS (Whelan et al. 1976).

Chevalier,C., Janot-Pacheco,E., Mauder,H., Ilovaisky,S.A.(1976)
    Paper presented at this meeting.
Whelan,J.A.J., Ward,M.J., Allen,D.A., Danziger,I.J., Fosbury,R.A.E.,
    Murdin,P.G., Penston,M.V., Peterson,B.A., Wampler,E.J., and
    Webster,B.L. (1976) Preprint.

A NOTE ON THE SPECTRUM OF V 616 MONOCEROTIS

H. W. Duerbeck

Sternwarte der Universitaet Bonn, F.R.G., and
European Southern Observatory, La Silla, Chile

Spectroscopic observations of the optical counterpart of the
X-ray nova A0620-00 (= V616 Mon) have been made by various observers with different equipment at most stages of its development.
Observations in August and September, 1975, show a practically
continous spectrum with weak emission lines of He II 4686 and N III
4640, while observations at later stages (early 1976) reveal the
spectrum of a cooler source, with absorption lines of He I and
possibly Fe II (Chiatti et al., this symposium). These findings
agree with photoelectric observations, which show a slow but steady
increase in colour index. Unlike in ordinary novae, the brightness
increase of V616 Mon can be explained by a temperature increase of
parts of the surface of the star, rather than by an increase of
surface area from an optically thick expanding envelope.

A Cassegrain spectrogram (IIa-O, 86 $\text{Å} \cdot \text{mm}^{-1}$) obtained on September 24, 1975 with the ESO 1 m telescope reveals several emission
features. He II 4686 is a relatively strong, broad line. The broad-
ening corresponds to a rotational velocity of $\pm$ 700 $\text{km} \cdot \text{s}^{-1}$. The
broad blend of N III, C III near 4640 is somewhat weaker and shows
a structured appearance. N III 4284 and He II 4542 are much weaker
and diffuse. Hβ , Hγ and Hϵ are weakly present, Hδ is somewhat
stronger, showing two sharp emission components separated by 1400
$\text{km} \cdot \text{s}^{-1}$, with no or weak absorption between them. The same structure
is suspected for the other H lines. The spectrum resembles that of
Sco X-1, with weaker H lines and much broader line profiles.

A crucial feature for future model construction is the broad
structure of the major emission lines. If due to rotation, the
lines cannot originate in the atmosphere of the normal star, but
possibly in an accretion disk surrounding the compact companion.

M. Friedjung (ed.), Novae and Related Stars, 150. All Rights Reserved.
Copyright © 1977 by D. Reidel Publishing Company, Dordrecht, Holland.

SUMMARY OF MORNING SESSION, SEPTEMBER 8

NEBULAR STAGE OF NOVAE; TRANSIENT X-RAY SOURCES

MUSTEL, E.R.

Astron. Council USSR, Acad. of Sci. Moscow, USSR.

The nebular stage is a very important stage in the evolution of every Nova after light maximum. This stage permits to study fine structure of principal envelopes ("polar caps", equatorial belts and so on) and to study the chemical composition of these envelopes, especially for those elements which have relatively high excitation and ionization potentials (for example He).

The chemical analysis of principal envelopes of Novae is connected with a solution of difficult problem - the problem of ionization and excitation of atoms in these envelopes. It seems that the envelopes of certain Novae contain regions which are characterized by widely different temperatures from thousands to millions degrees, a collisional excitation seems to be very important. A study of the origin of "coronal" emission in the spectra of Novae (localization of emission and the source of high temperature particles which are responsible for this emission) is a very important problem.

The results of several studies of the chemical analysis of envelopes of Novae indicate that the relative abundances of metals and the hydrogen-metal ratio are approximately the same as the solar abundances. There are overabundances of C, N, O; for C this overabundance is the least. The relative abundance of helium is determined not very well. The task to study the energy distribution in the spectra of Novae after light maximum in the UV and X-ray regions is also very important.

A new very interesting aspect of the analysis of Novae are observations of transient X-sources. Some of these sources belong to Novae or to Nova-like objects

M. Friedjung (ed.), Novae and Related Stars, 151-152. All Rights Reserved.
Copyright © 1977 by D. Reidel Publishing Company, Dordrecht, Holland.

and the observations of all these objects may give very
important information about the physics of Novae them-
selves.

PART IV

OBSERVATIONS OF NOVA CYGNI 1975

OBSERVATIONS DE NOVA CYGNI 1975 (V 1500 Cyg) [*]

ANDRILLAT Yvette

Observatoire de Haute Provence - France

Laboratoire d'Astronomie USTL

34060 Montpellier Cédex - France

Sommaire

Nova Cyg 1975 est la plus brillante nova apparue depuis 33 ans. Elle a subi une variation d'éclat supérieure à 19 magnitudes, jamais observée jusque là. C'est une nova extrêmement rapide. Après le maximum, son déclin est caractérisé par une diminution d'éclat de 3,9 magnitudes en 3 jours. Sa distance est estimée à 1300 parsecs environ.

De rapides variations de brillance de période 3,3 heures environ ont été découvertes. Un brusque changement dans la répartition énergétique spectrale du continu a été enregistré. Au maximum et 3,2 jours après, cette répartition est analogue à celle d'un corps noir. Ensuite, elle ressemble à celle d'une source de rayonnement "free-free".

L'évolution spectrale a été très rapide, le stade nébulaire étant atteint 13 jours seulement après le maximum. Pendant la

[*] Les observations ont été effectuées en partie à l'Observatoire de Haute Provence (CNRS).

M. Friedjung (ed.), Novae and Related Stars, 155-176. All Rights Reserved.
Copyright © 1977 by D. Reidel Publishing Company, Dordrecht, Holland.

première phase du déclin, le spectre a été caractérisé par
l'absence de systèmes d'absorption diffus, diffus renforcé et
d'Orion. Aucune nébulosité n'a encore été observée autour de
la nova.

Les spectres rouge et infrarouge jusqu'à 1,1μ montrant en
particulier les raies coronales de $[FeX]$ 6374 Å, $[FeXI]$ 7892 Å,
$[SVIII]$ 9911 Å confirment les identifications de raies coronales
de Al, Si, Ca, à différents degrés d'ionisation, faites dans la
région 2 à 3 microns.

———————

V 1500 Cyg est la nova la plus brillante apparue depuis
CP Pup 1942. A son maximum d'éclat, le 30,7 août 1975, elle
a atteint la magnitude visuelle 1,7 $\pm$ 0,5.

Aucune étoile n'a été trouvée à son emplacement sur le Sky
Atlas du Mt Palomar, malgré les mesures précises de position
de la nova (IAU Circ. 2826, 2829, 2837) et les recherches
effectuées sur des photographies obtenues, après sa découverte,
par Beardsley et al (1975).

V 1500 Cyg est donc caractérisée par une augmentation
d'éclat $\Delta m > 19$ magnitudes qui est la plus grande variation de
brillance observée pour une nova galactique. Dans le cas de
CP Pup : $\Delta m \# 17$.

V 1500 Cyg a fait l'objet de très nombreuses observations.
Les déterminations de magnitudes visuelles sont résumées dans :
IAU Circ. 2826, 2830, 2832, 2842, 2846, 2848, 2851, 2857, 2873,
2885, 2902, 2914, 2938, 2953, 2973 (Bocek et al, 1976). Par
ailleurs, la nova a été retrouvée sur des photographies prises
avant sa découverte, notamment sur des clichés obtenus en
août le 12,0 TU ($m_{pg} = 16$), le 28,059 TU ($m_{pv} > 9,6$), le 29,052
($m_{pv} = 7,5$)... IAU Circ. 2826, 2839, 2848, 2864. De plus, des

mesures de magnitudes ont été faites à travers différents filtres: U, B, V, R, I, J, H et K, et des indices de couleur ont été mesurés (IAU Circ. 2826, 2828, 2829, 2830, 2832, 2834, 2839, 2846, 2853, 2857, 2858, 2864), (Rosino et Tempesti, 1976), (Marcocci et al, 1976), (Gallagher et Ney, 1976), (Lindegren et Lindegren, 1975), (Ichimura et al, 1975).

La courbe de lumière déduite de ces données montre une forme exponentielle après le maximum, ce dernier étant arrondi et suivi d'un déclin très rapide de 3 magnitudes environ en 3,9 jours (t_3 = 3,9 jours, t_3 est le nombre de jours comptés à partir du maximum, correspondant à une diminution d'éclat de 3 magnitudes. Pour CP Pup, t_3 était égal à 7 jours).

La grande variation d'éclat ($\Delta m > 19$ magnitudes), la forme exponentielle de la courbe de lumière avec son maximum arrondi suggèrent qu'il s'agit d'une supernova de type I, mais les caractéristiques spectrales, les vitesses d'expansion sont typiques d'une nova. D'ailleurs, pour une supernova, la diminution d'éclat à partir du maximum est environ 5 fois plus lente.

Pour une nova, il existe différentes relations entre la variation de sa luminosité au cours du temps, après le maximum, et la magnitude absolue visuelle M_o au maximum :

$$M_o = -11,5 + 2,5 \log t_3 \text{ (McLaughlin, 1960) donne } M_o = -10,0$$

A partir de l'étude de 27 novae extragalactiques et 11 novae galactiques, Pfau (1976) a établi une relation plus précises :

$$M_{oB} = -10,67 \ (\pm 0,30) + 1,80 (\pm 0,20) \log t_3$$

M_{oB} étant la magnitude absolue photographique au maximum.
On trouve $M_{oB} = -9,61$.

La rapidité du déclin au cours du temps, calculée en extra-
polant la courbe de lumière, a conduit De Vaucouleurs (IAU
Circ. 2839) à $M_o = -10,25$.

Pour accéder à la distance de la nova, l'absorption interstellaire a été étudiée par différentes méthodes :

V 1500 Cyg est située dans le plan galactique, à la limite de
deux régions où A_V varie de 0,7 à 3,0 magnitudes par kiloparsec.
En adoptant la relation

$$m_o - M_o = 10 + 5 \log r + r A_V$$

où r est la distance en kpc et m_o la magnitude apparente au
maximum, on constate que r varie de 1400 à 800 parsecs
(Ichimura et al, 1975).

La connaissance de l'indice de couleur des novae au maximum d'éclat $(B-V)_o = 0,35 \pm 0,05$ magnitudes (Schmidt, 1957)
conduit dans le cas de V 1500 Cyg, à une valeur de l'extinction
interstellaire de $A_V = 0,8$ ou $A_V = 1,6$ magnitudes, soit à des
distances respectives de 1850 et 1150 pc (Pfau, 1976).

Une étude de l'excès de couleur en fonction du module de
distance pour une vingtaine d'étoiles situées dans un champ de
3° autour de la nova a conduit Woszczyk et al (1975) à une
distance $830 < d < 1100$ pc.

Récemment l'observation de nombreuses étoiles dans la
direction et au voisinage de la nova a permis de déduire la loi de
rougissement en fonction de la distance (Schild, 1976) l'excès
de couleur $E(B-V) = 0,12$ magnitude et la distance $d < 1500$ pc.

La valeur du rougissement est en bon accord avec celle obtenue
par les observations dans le proche ultraviolet par le satellite
Copernicus (Jenkins et al, 1976).

Le spectre de V 1500 Cyg. présente plusieurs raies inter-
stellaires notamment celles de CaII 3969 $\overset{\circ}{A}$, CaI 4227 $\overset{\circ}{A}$,
CH^+ 4232 $\overset{\circ}{A}$, NaI 5890 $\overset{\circ}{A}$, Li 6708 $\overset{\circ}{A}$, KI 7699 $\overset{\circ}{A}$. La largeur
équivalente et le profil de ces raies, mesurés sur des "scans"
photoélectriques à haute résolution 0,03 à 0,10 $\overset{\circ}{A}$, ont été
comparés à 55 Cyg. On en conclut que la nova est à une distance
plus grande que 55 Cyg et $- 10,5 < M_o < - 9,5$ (Tomkin et al,
1976).

A partir des intensités de ces raies interstellaires De
Vaucouleurs (IAU Circ. 2839) trouve une distance d = 1300
$\pm$ 200 parsecs.

La lumière de la nova est polarisée : la valeur moyenne
trouvée, 1,20 %, correspond à la polarisation interstellaire
dans la région du Cygne. McLean (1976) a observé une polari-
sation de 1,38 $\pm$ 0,04 % à 5460 $\overset{\circ}{A}$. Il l'attribue principalement à
une diffusion de la lumière par le milieu interstellaire. Il n'ex-
clut pas la présence d'une polarisation intrinsèque de la lumière
de la nova, mais il l'estime inférieure à 0,18 %. Quelques
légères variations du taux de polarisation ont été observées : la
plus intense, d'amplitude 0,5 % et de période 14 jours, a été
enregistrée entre le 27 septembre et le 7 novembre (IAU Circ
2828, 2829, 2832, 2834, 2837, 2873, 2973).

La polarisation interstellaire donne une estimation grossière
de la distance 828 < d < 1445 pc et de la magnitude
$- 10,65 < M_o < -9,44$.

Enfin, une méthode indirecte a fourni une valeur de la
distance à partir des mesures photométriques infrarouges à

large bande dans le domaine $0,5 - 10\mu$.

Au voisinage du maximum, la répartition énergétique spectrale dans le continu est approximativement celle d'un corps noir : la température de couleur et la brillance de surface sont donc définies. De plus, la quantité de flux reçue, les vitesses radiales et les vitesses d'expansion sont mesurées. On en déduit le rayon angulaire de la source et la distance $1200 < d < 2300$ pc (Gallagher et Ney, 1976).

Ces diverses déterminations de M_o et de d conduisent à une valeur moyenne voisine de $M_o = - 10,1$ et $d = 1300$ pc. La magnitude absolue de la prénova est de $+ 10,5$ environ (Starrfield et al, 1976), valeur typique pour une naine blanche. Cependant, les prénovae classiques sont généralement une cinquantaine de fois plus brillantes. La grande et inhabituelle augmentation d'éclat de V 1500 Cyg pourrait ainsi s'expliquer par la luminosité anormalement faible de la prénova.

Le rayonnement de Nova Cygni 1975 a été étudié dans un très vaste intervalle spectral.

Dans le domaine radio, aucune émission n'a été détectée à la fréquence de 10,6 GHz au moment du maximum d'éclat, la limite inférieure du flux détectable étant de 10 mJy. De même, jusqu'au 8 octobre, à la fréquence de 2,655 GHz, aucun flux supérieur à 4 mJy n'a été décelé. Par contre, à 8,085 GHz, une émission de 6 ± 2 mJy a été enregistrée le 24 septembre. Elle était de 12 mJy le 1er octobre et de 14 mJy le 8 octobre. Une augmentation analogue a été constatée à la fréquence de 10,7 GHz: la valeur du flux était de 10 ± 5 mJy le 27 septembre, et de 19 ± 5 mJy le 8 octobre (IAU Circ. 2826 - 2851 - 2853).

Le 4 septembre, une explosion a été observée à 989 MHz

(Altunin, 1976).

Quant au rayonnement X de la nova, il est très faible :

dans l'intervalle d'énergie 2 - 20 keV, le rapport du rayonnement

X au rayonnement visible est estimé 3.10^7 fois plus faible que

celui trouvé pour la source X intermittente A0 620-00 (IAU Circ.

2828 - 2829).

Le comportement de V 1500 Cyg dans l'infrarouge s'est

révélé fort intéressant.

Les mesures photométriques à larges bandes effectuées jusqu'à

$1,25\mu$ durant 50 jours après le maximum montrent un important

et brusque changement de la répartition énergétique spectrale

dans le continu (Gallagher et Ney, 1976) : durant les 3 premiers

jours après le maximum, cette répartition est pratiquement

celle d'un corps noir de température 5000°K. Après le quatriè-

me jour, elle ressemble à celle d'une source de rayonnement

"free-free" : elle est analogue à celle de l'émission de Nova RR

Tel qui a évolué fort lentement. Ce rapide changement suggère

qu'au début de l'expansion, l'enveloppe est optiquement épaisse

pendant 3,2 jours. Les calculs montrent que la masse éjectée

est de l'ordre de 10^{-4} à 10^{-5} $M_\odot$ et que l'énergie cinétique est

supérieure à 10^{44} ergs, valeurs en bon accord avec celles

trouvées pour les novae galactiques.

Dans la région bleu-visible, une rapide variation de

brillance a été découverte le 9 septembre (IAU Circ.2834). La

période 0,14096 jours, est constante tandis que l'amplitude est

variable. Une variation analogue a été observée le 18 septembre,

ainsi qu'un minimum secondaire dont l'intensité varie avec la

longueur d'onde (Marcocci et al, 1976). Ces variations ont été à

nouveau enregistrées en juin et juillet 1976 par Kemp et al

(IAU Circ. 2973 - 2981) dans l'intervalle 3500-5500 Å, avec

une période de 0,1384 jours, en bon accord avec la valeur

précédente, et une amplitude de 0,20 magnitude. Ces auteurs

ont également noté de très rapides variations d'éclat en 1 à 3

minutes, l'amplitude étant de 0,027 magnitude.

Ces variations s'interprètent (Rosino, Tempesti, 1976) en

supposant que V 1500 Cyg est une étoile double : en effet la

période étant très courte, 3,3 heures, les 2 étoiles sont très

proches et après l'explosion, l'étoile orbitante se trouve à

l'intérieur de l'enveloppe en expansion, créant une perturbation

qui peut se traduire par une variation périodique de brillance.

Toutefois, cette binarité est discutée par Starrfield et al (1976)

qui suggèrent que le comportement de Nova Cygni 1975 s'expli-

que par une explosion thermonucléaire survenant dans l'enve-

loppe mince d'une naine blanche isolée.

V 1500 Cyg a fait l'objet de très nombreuses observations

spectrographiques de l'ultraviolet à l'infrarouge avec des

grandes, des moyennes et des faibles résolutions (IAU Circ.

2827, 2829, 2832, 2834, 2839, 2842, 2857) (Campbell, 1976)

(Fehrenbach et Andrillat, 1975) (Fehrenbach, Andrillat, 1976)

(Grasdalen et Joyce, 1976) (Ichimura et al, 1976) (Leparkas,

1976) (Rosino et Tempesti, 1976) (Sigal et Stal'bovsky, 1976)

(Tomkin et al, 1976) (Woszczyk et al, 1976). Je me bornerai à

signaler les résultats les plus saillants qui ont été obtenus en

développant davantage ceux relatifs à la région du proche infra-

rouge où peu de novae ont été observées.

Quelques spectres ont pu être pris avant le maximum. Ils

sont caractérisés par un continuum intense sur lequel on distingue difficilement quelques absorptions d'origine stellaire. Dans la région du proche infrarouge (figure 1) (Fehrenbach et Andrillat, 1975), le 30,06 août (TU), seule celle de OI 7772 $\overset{\circ}{A}$ est bien visible alors que H_α se devine à peine. A partir du 30,95 août, le continu devient moins intense et de larges émissions avec un profil P Cyg sont visibles. La largeur des absorptions montre l'existence d'un gradient de vitesses dans l'enveloppe où elles prennent naissance. La vitesse radiale déduite de ces absorptions est de l'ordre de − 1300 km.s^{-1} le 30,06 août, − 1700 km.s^{-1} le 30,95 août et de − 1900 km.s^{-1} le 2,90 septembre.

Au maximum, le spectre d'absorption est analogue à celui d'une supergéante de type A; l'indice de couleur est également celui d'une telle étoile (IAU Circ. 2829).

Jusqu'au 2,90 septembre, on note le développement des raies de l'hydrogène de la série de Paschen, de celles de NI et de OI 7772 et 8446 $\overset{\circ}{A}$ (figure 1). Ces spectres, de dispersion 31 $\overset{\circ}{A}$.mm^{-1} ont été obtenus au spectrographe du foyer coudé du télescope de 152 cm de l'Observatoire de Haute Provence.

Avec ce même spectrographe, mais avec une dispersion de 12 $\overset{\circ}{A}$.mm^{-1}, nous avons étudié la région 3600–5050 $\overset{\circ}{A}$. La figure 2 présente les spectres des 31,1 août et 2,8 septembre. Outre de nombreuses raies de FeII et TiII, celles de l'hydrogène de la série de Balmer sont visibles jusqu'à H_{10} et deviennent ensuite très intenses. Par contre, les observations faites avec Copernicus ne montrent pas L_α, peut-être à cause de l'absorption interstellaire ou bien à cause d'une capture possible de la raie par la nébulosité.

Dans l'évolution spectrale des novae, on distingue généra-

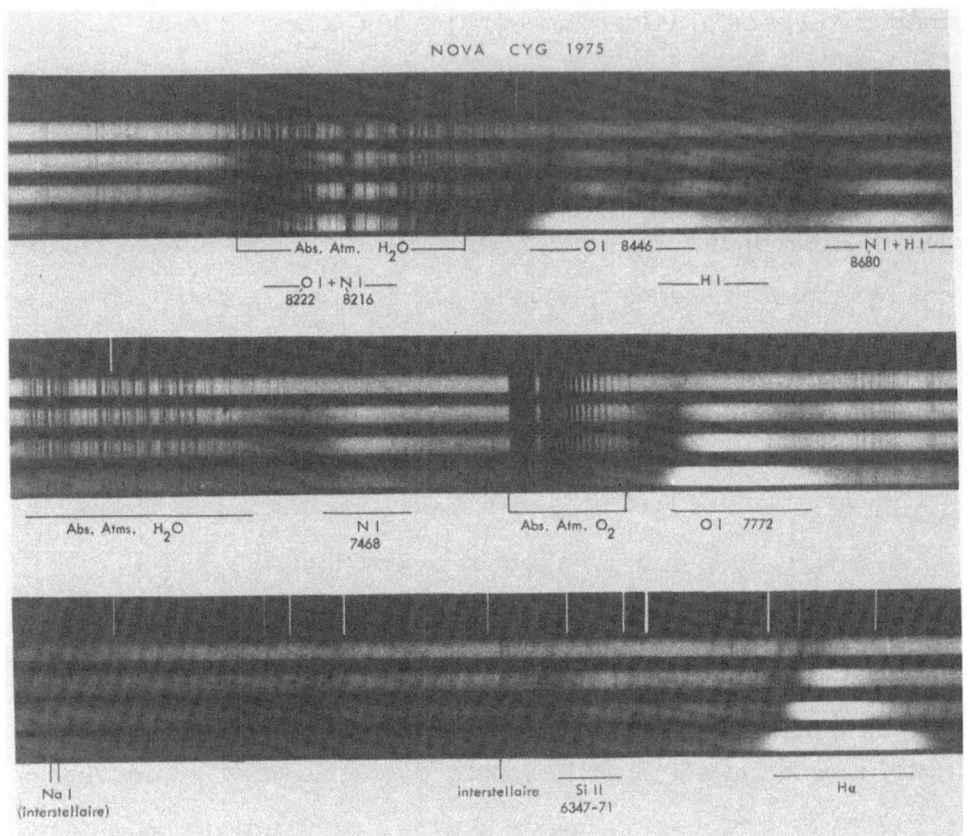

Figure 1-

Portions de 4 spectres de V 1500 Cyg dans la région du proche infrarouge, de 5800 à 8750 Å, obtenus à l'aide du spectrographe du foyer coudé du télescope de 152 cm de l'Observatoire de Haute Provence, sur plaques Eastman Kodak IN hypersensibilisées à l'ammoniaque.

La dispersion originale est de 31 Å.mm^{-1}.

De haut en bas, spectres obtenus le 30,06 août (TU)

le 30,95 août (TU)

le 31,97 août (TU)

le 2,90 septembre (TU).

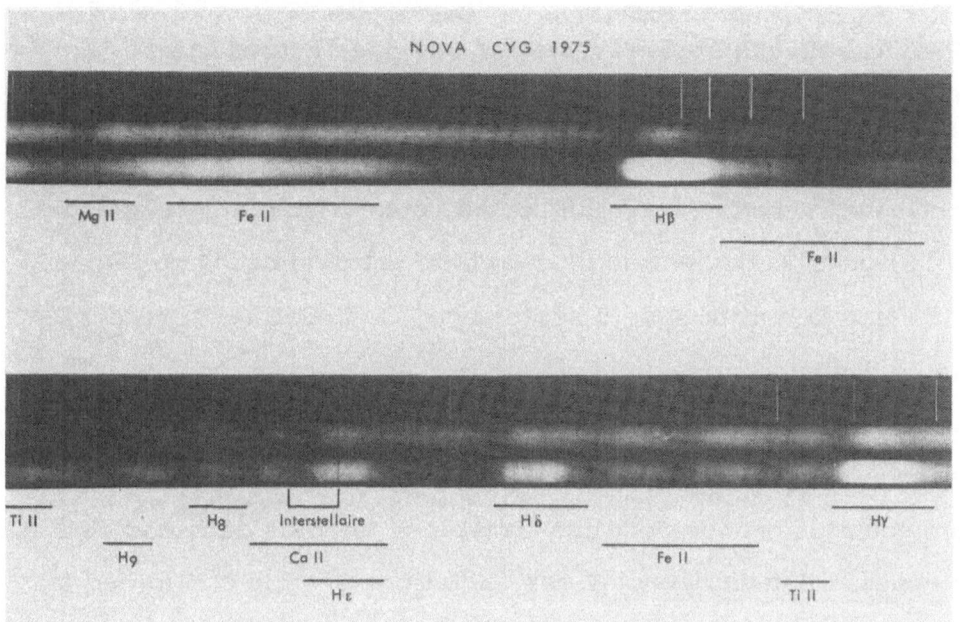

Figure 2–

Deux spectres de V 1500 Cyg dans la région 3600–5050 Å, obtenus au spectrographe du foyer coudé du télescope de 152 cm de l'Observatoire de Haute Provence sur plaques Eastman Kodak IIaO chauffées.

La dispersion originale est de 12 Å.mm^{-1}.

De haut en bas, spectres pris le 31,1 août (TU)

le 2,8 septembre (TU).

Sur le spectre du 2,8 septembre, le continu est plus faible et les émissions plus intenses et plus larges. On remarque leur structure complexe bien visible notamment pour la raie H$_\delta$ où l'on distingue nettement 4 maximums d'intensité.

lement différents systèmes d'absorption bien connus : diffus, diffus renforcé et spectre d'Orion. Dans le cas de Nova Cygni 1975, ces divers stades n'existent pas : les larges raies d'absorption ne montrent aucune structure.

Quant aux raies d'émission, à partir du 2 septembre, elles sont très intenses et leur structure, bien visible sur le spectre de la figure 2, devient de plus en plus nette au cours du temps.

Dans la nuit du 4 au 5 septembre, Campbell (1976) a enregistré des variations du profil de Hα en quelques heures. Il les attribue d'une part à des fluctuations dans la source centrale de rayonnement ionisant, d'autre part à la durée du trajet de la lumière plus grande pour les émissions les plus décalées vers le rouge. L'étude détaillée des variations semble confirmer la nature binaire de la nova.

Les émissions montrent 4 composantes principales d'intensité variable au cours du temps. Les figures 3 et 4 donnent les profils de Hγ et Hδ les 4, 6, 8 septembre et 22 octobre. Les vitesses radiales des différentes composantes sont sensiblement les mêmes et on peut admettre que, le 22 octobre, l'enveloppe a atteint une certaine stabilité de structure. Les différents pics d'intensité peuvent être interprétés comme des condensations de gaz se déplaçant radialement dans diverses directions. L'enveloppe serait alors formée de zones distinctes : un modèle composé de 2 calottes polaires et de 2 anneaux équatoriaux serait compatible avec nos observations, la ligne de visée formant un angle d'une dizaine de degrés avec le plan équatorial (Fehrenbach et Andrillat, 1976).

Dans le cas d'un modèle à symétrie grossièrement sphérique, avec les valeurs trouvées pour la vitesse d'expansion,

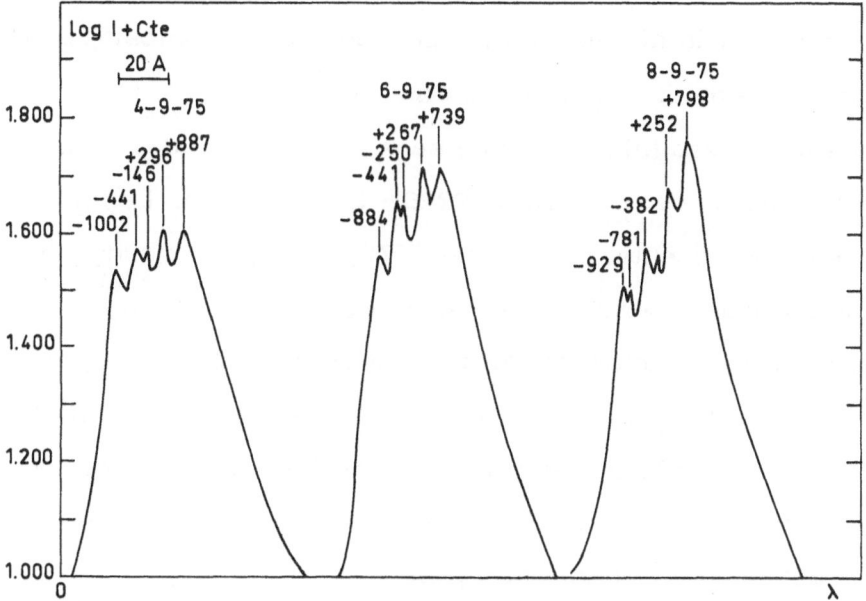

Figure 3- Profil photométrique de la raie Hγ les 4, 6 et 8 septembre 1975. Celui du 22 octobre n'est pas reproduit car, à cette date, la raie est mélangée avec [OIII] 4363 Å. Dispersion originale : 20 Å.mm^{-1}.

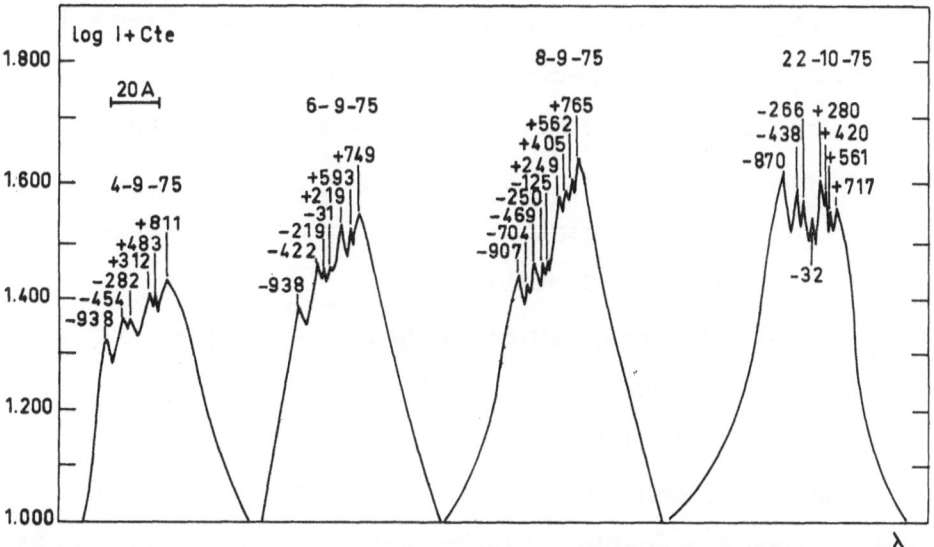

Figure 4- Profil photométrique de la raie Hδ . La présence de [SII] 4068-4076 Å sur l'aile violette de la raie donne un aspect fortement asymétrique à ce profil. Dispersion originale : 20 Å.mm^{-1}.

– 1300 km.s^{-1} et la distance, 1400 parsecs, une nébulosité d'un diamètre 0''4 devrait actuellement être visible.

Plusieurs tentatives faites dans ce sens aboutirent à des conclusions contradictoires (IAU Circ. 2926 – 2938 – 2953). Les observations les plus récentes, en juillet 1976, faites par Young par excellentes images infirment ce résultat (IAU Circ. 2981).

L'évolution spectrale de Nova Cygni 1975 a été très rapide puisque le stade nébulaire était atteint le 12 septembre, soit 13 jours seulement après le maximum. A partir de cette date, les spectres présentent les raies permises et interdites caractéristiques des novae pendant cette phase.

La figure 5 montre les spectres obtenus à l'Observatoire de Haute Provence, au télescope de 120 cm avec une dispersion de 230 Å.mm^{-1} dans la région 5700–8750 Å (Fehrenbach et Andrillat, 1976). Du 2 octobre au 10 novembre, Hα et OI 8446 Å dominent les spectres, puis OI diminue pour disparaître complètement à partir du 7 janvier. Le spectre de HeI est bien développé (7065 Å, très intense, 6678 et 5875 Å). NI est visible jusqu'au 10 novembre.

Les raies interdites augmentent régulièrement d'intensité d'octobre à janvier. On identifie :

– [NII] 5755 Å : la surexposition de Hα ne permet pas de distinguer 6548 et 6584 Å sur ses ailes

– [FeVII] 6086 Å est visible à partir du 10 novembre. La raie 5721 Å, dont le niveau supérieur est le même que celui de 6086 Å, est également présente. Nous mesurons sur le spectre du 25 novembre un rapport d'intensité I(6086)/I(5721) égal à 1,4, en bon accord avec la valeur théorique 1,58. On en conclut que si la raie [CaV] 6085 Å est présente, sa contribution est très

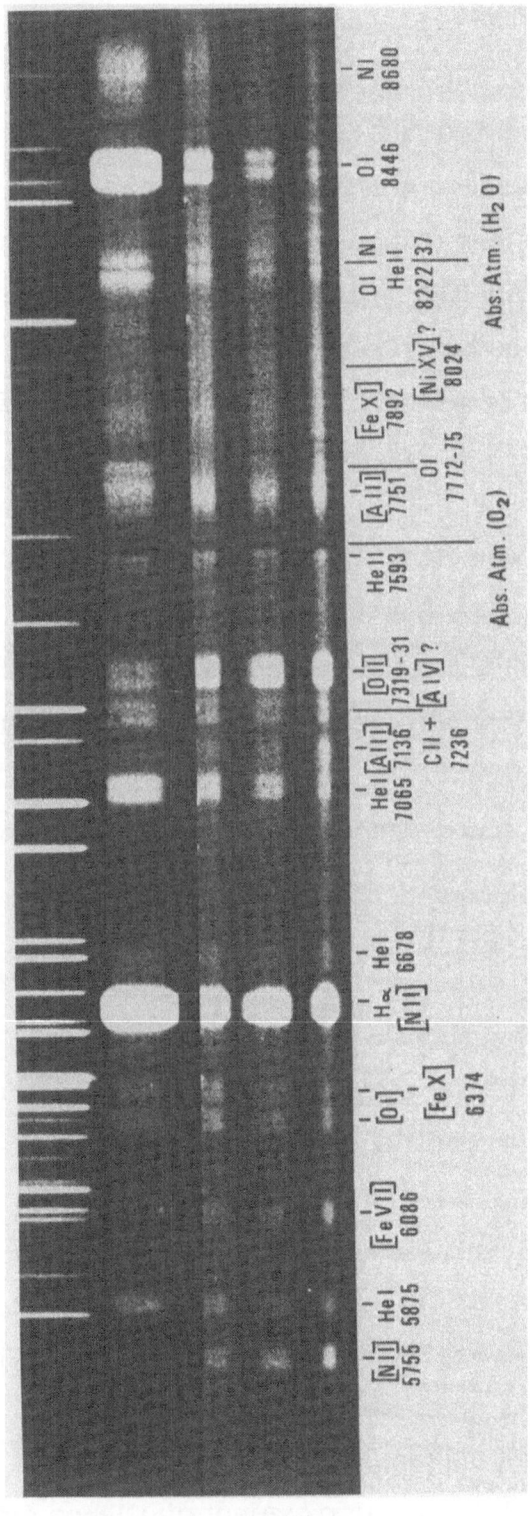

Figure 5– Spectres de V 1500 Cyg obtenus les 2–10–1975, 10–11–1975, 25–11–1975 et 7–1–1976, à l'aide du spectrographe E' au foyer Newton du télescope de 120 cm de l'Observatoire de Haute Provence sur plaques Eastman Kodak IN hypersensibilisées à l'ammoniaque. Dispersion originale : 230 Å.mm^{-1}. Les émulsions qui débordent le dernier spectre sont dues aux raies de OH du ciel nocturne.

peu importante.

- $[OI]$ 6300 et 6363 Å. La variation du rapport d'intensité de ces 2 raies est importante : 1,0 le 2 octobre, 1,10 le 10 novembre, 1,40 le 25 novembre et 2,26 le 7 janvier. Ces valeurs inférieures à la valeur théorique 3,14 peuvent s'expliquer par le mélange de $[OI]$ 6363 Å avec la raie coronale de $[FeX]$ 6374 Å. La présence de cette raie à cette époque est vraisemblable puisque de nombreuses raies coronales ont été observées dans la région 2-3 μ à partir du 29 septembre (Grasdalen et Joyce, 1976). La présence de SiII 6348 Å mélangée à $[OI]$ 6363 Å est certaine à la fin du mois d'août (IAU Circ. 2829) et probable en septembre. Ensuite, on note une forte diminution de l'intensité des raies ayant un faible potentiel d'ionisation : en octobre l'intensité de SiII doit être très faible ou nulle.

- $[AIII]$ 7136 Å et 7751 Å, cette dernière étant mélangée avec OI 7772 Å.

- $[OII]$ 7319-7331 Å qui, à partir du 25 novembre domine le spectre avec Hα

- $[FeXI]$ 7892 Å dont la longueur d'onde est voisine de MgII 7877-7896 Å. La présence de cet élément est probable jusqu'au début septembre, période pendant laquelle le doublet 4481 Å est visible.

Les figures 6 et 7 montrent les spectres obtenus au foyer Cassegrain du télescope de 193 cm de l'Observatoire de Haute Provence dans la région 7600-11300 Å avec un tube image ITT à photocathode S1. Quelques raies de CI et SI sont encore visibles sur le spectre du 24 septembre. Plusieurs raies d'éléments à faible potentiel d'ionisation avaient d'ailleurs été

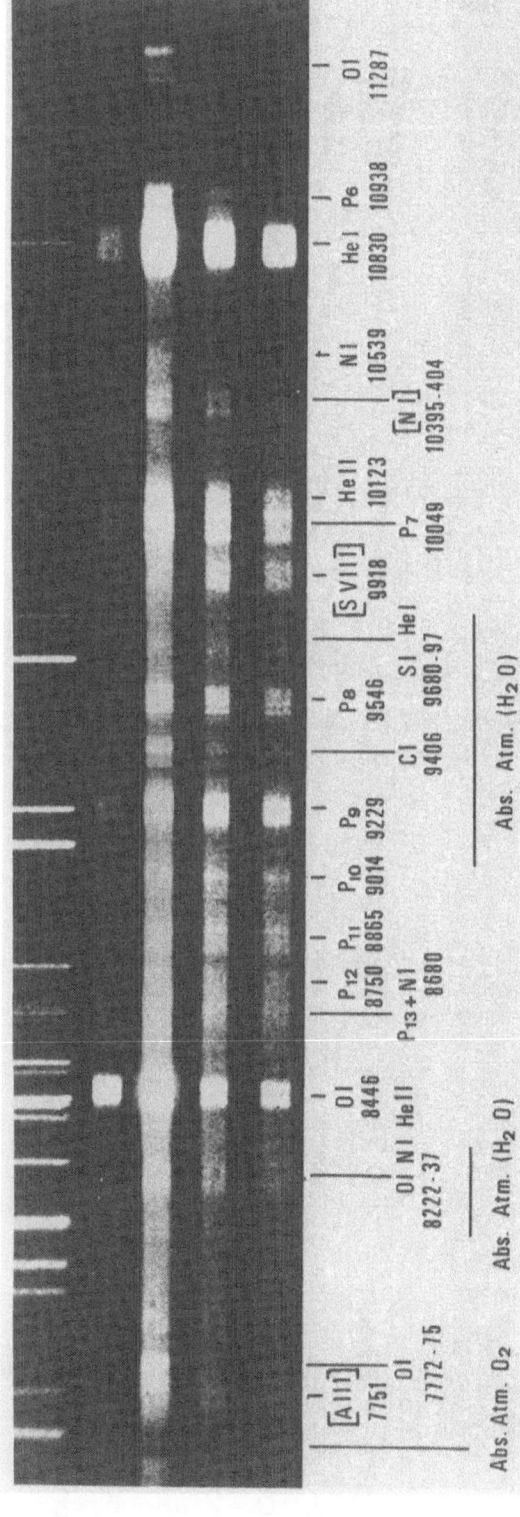

Figure 6– Spectres de V 1500 Cyg dans la région 0,77–1,13 microns obtenus les 22–9–1975, 24–9–1975, 10–11–1975 et 13–11–1975, à l'aide du spectrographe ROUCAS au foyer Cassegrain du télescope de 193 cm de l'Observatoire de Haute Provence. Le récepteur est un tube image ITT F4718 à 2 étages à photocathode S1. La dispersion originale est de 230 Å.mm^{-1}. L'émulsion utilisée est du film Eastman Kodak 103aD.

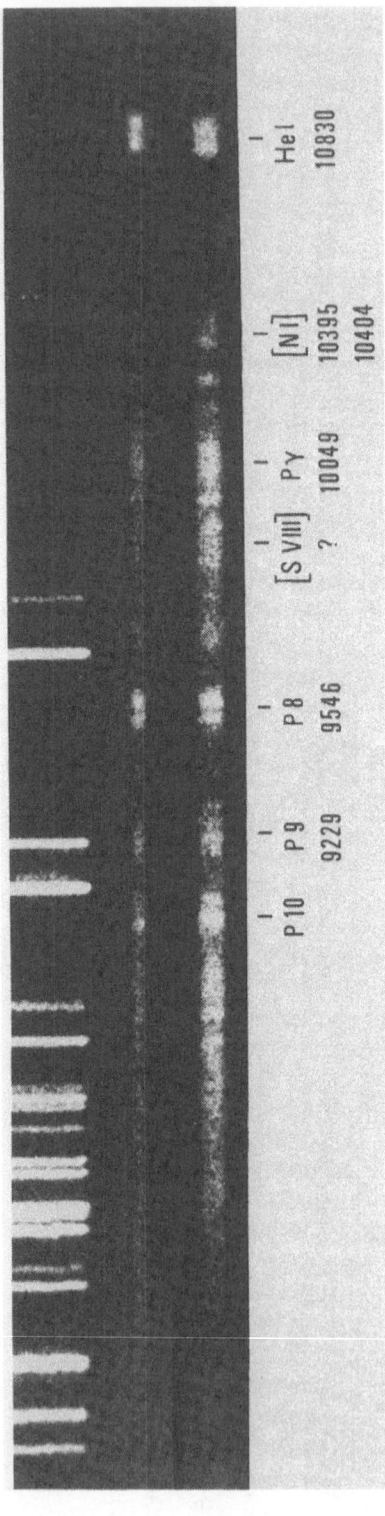

Figure 7– Spectres de V 1500 Cyg dans la région 0,77–1,13 microns, obtenus les 15–6–1976 et 30–7–1976 à l'aide du spectrographe ROUCAS au foyer Cassegrain du télescope de 193 cm de l'Observatoire de Haute Provence. Le récepteur est un tube image ITT F4718 à 2 étages et à photocathode S1. La dispersion originale est de 230 Å.mm^{-1}. L'émulsion utilisée est du film Eastman Kodak 103aD.

Seules les raies de l'hydrogène de la série de Paschen P_γ, P_8, P_9, P_{10} et la raie de HeI à 10830 Å sont intenses.

préalablement identifiées dans ce domaine spectral dans les
spectres de Nova Scuti 1975 obtenus après le maximum d'éclat
(Fehrenbach et Andrillat, 1975). Sur nos spectres, CI 9406 Å,
SI 9680 Å, NI 10539 Å diminuent fortement d'intensité à partir
du 24 septembre tandis que les raies de la série de Paschen se
détachent nettement du continu. HeI 10830 Å domine nettement
le spectre.

Sur le spectre surexposé du 24 septembre on note la présence
de OI 11287 Å. Elle a été également trouvée après le 2 septembre (Tomkin et al, 1976) et le 30 septembre (IAU Circ.2842) par
transformée de Fourier alors que 13164 Å est absente. Ces
observations sont importantes car elles confirment que 8446 Å
est bien formée par un mécanisme de fluorescence à partir de
L_β , explication donnée lors d'études de novae précédentes.
Cette région spectrale est pauvre en raies interdites.
[NI] 10395-10404 Å est encore visible, mais faible sur le spectre du 30 juillet 1976 caractérisé par les fortes émissions de
HeI 10830 Å, P_γ , P8, P9 et P10 (figure 7).

Une intense émission visible sur tous les spectres a été
mesurée à 9913 $\overset{+}{-}$ 5 Å. Nous l'avons attribuée à [SVIII] 9911 Å
qui a été trouvée par Kissel et Byard (1965) lors de l'éclipse de
soleil en 1965. On la devine encore bien que très faible, sur le
spectre du 30 juillet.

Dans la région 5700-11000 Å le spectre de V 1500 Cyg est
caractérisé par les raies coronales de [FeX] 6374 Å,
[FeXI] 7892 Å et [SVIII] 9911 Å, dont la présence n'a jamais
été signalée à notre connaissance dans le spectre d'une nova.

Ces identifications sont importantes car elles complètent
les observations des raies coronales faites par Grasdalen et
Joyce (1976) dans la région de 2-3μ , de [SiVI] , [SiVII] ,

$\left[\text{SiIX}\right]$, $\left[\text{AlV}\right]$, $\left[\text{AlVI}\right]$, $\left[\text{AlVIII}\right]$, $\left[\text{AlIX}\right]$, $\left[\text{CaIV}\right]$, $\left[\text{CaVI}\right]$

et $\left[\text{MgVIII}\right]$. Ces auteurs signalaient par ailleurs l'absence des

raies coronales dans les autres régions spectrales.

Les identifications des raies coronales proposées par
Grasdalen et Joyce sont discutées par Black et Gallagher (1976)
qui proposent de les attribuer à des raies produites par un mécanisme de recombination et appartenant aux triplets de HeI. Nos
observations semblent confirmer la présence des raies coronales dans le domaine spectral 2-3μ . Grasdalen et Joyce enre-
gistrent une augmentation de leur intensité au cours du temps :
il se développe donc à l'intérieur du matériel éjecté par la nova
une région à très haute température $T \sim 10^{6}$°K.

De plus ces raies sont dues aux éléments Al, Si, Ca ionisés
à des degrés différents et la mesure de leurs intensités devrait
donc permettre de déduire les conditions physiques régnant à
l'intérieur de la zone coronale.

Bibliographie

Altunin, V.I. 1976, Astron. Zh., 2, 299.

Beardsley, W.R., King, M.W., Russell, J.L. and Stein, J.W.
        1975, Publ. Astron. Soc. Pacific, 87, 943.

Black, J.H. and Gallagher, J.S. 1976, Nature, 261, 296.

Boček, J., Ceplecha, Z., Ježková, M. and Novák, M. 1976,
        Bull. Astron. Inst. Czech., 27, 190.

Campbell, B. 1976, Astrophys. J., 207, L.41.

Fehrenbach, Ch. et Andrillat, Y. 1975, Compt. Rend. Acad. Sci.
        Paris, 281, 189.

Fehrenbach, Ch. et Andrillat, Y. 1975, Compt. Rend. Acad.
        Sci. Paris, 281, 365.

Fehrenbach, Ch. et Andrillat, Y. 1976, Astron. Astrophys.,
    (sous presse).

Gallagher, J.S. and Ney, E.P. 1976, Astrophys. J., 204,
    L35.

Grasdalen, G.L. and Joyce, R.R. 1976, Nature, 259, 187.

IAU Circ. 2826, 2827, 2828, 2829, 2830, 2832, 2834, 2837,
    2839, 2842, 2846, 2848, 2851, 2853, 2857, 2858, 2864,
    2873, 2885, 2892, 2902, 2914, 2926, 2938, 2953, 2973,
    2981.

Ichimura, K., Nakagiri, M., Watanabe, E., Okida, K.,
    Nishimura, S. and Yamashita, Y. 1975, Tokyo Astron.
    Bull., 241, 2055.

Ichimura, K., Noguchi, T., Norimoto, Y. and Nariai, K. 1976
    Tokyo Astron. Bull., 242, 2061.

Jenkins, E., Snow, T., Upson, W., Starrfield, S., Friedjung,
    M., Gallagher, J.S., Linsky, J.L., Anderson, R., Henry,
    R.C. and Moos, H.W. 1976 (en préparation).

Kissell, K.E., Byard, P.L. 1965, Proc. 1965 Solar eclipse
    Symposium NASA Areas Research Center, 359.

Leparskas, H.J.Q. 1976, Publ. Astron. Soc. Pacific, 88, 154.

Lindegran, L., Lindegren, H. 1975, Nature, 258, 501.

McLaughlin, D.B. 1960, "The spectra of Novae" Stars and
    Stellar systems VI, Ed. Greenstein, Univ. of Chicago Press,
    Chicago.

McLean, I.S. 1976, Monthly Notices Roy. Astron. Soc., 176, 73.

Marcocci, M., Messi, R., Natali, G. and Rossi, L. 1976,
    Nature, 259, 186.

Pfau, W. 1976, Astron. Astrophys., 50, 113.

Rosino, L., Tempesti, P. 1976, Oss. Astrofis. di Asiago
    Preprint series n°1.

Schild, R.E. 1976, Center for Astrophysics Cambridge Mass.
      Preprint series n° 564.

Schmidt, Th. 1957, Z. Astrophys., 41, 182.

Sigal, G.P. and Stal'bovskij, O.I. 1976, Astron. Zh., 2, 303.

Starrfield, S., Truran, J.W., Gallagher, J.S., Sparks, W.M.,
      Strittmatter, P., Van Horn, H.M. 1976, Astrophys. J.,
      208, L23.

Tomkin , J., Woodman, J. and Lambert, D.L. 1976, Astron.
      Astrophys., 48, 319.

Woszczyk, A., Krawczyk, S., Strobel, A. 1975, Comm. 27
      IAU Inf. Bull. variable stars, décembre.

SPECTROPHOTOMETRIC STUDY OF THE CONTINUOUS SPECTRUM
OF NOVA CYGNI 1975 (V 1500 CYG)

C.T. HUA   Laboratoire d'Astronomie Spatiale du CNRS

N.H. DOAN   Observatoire de Lyon, Saint-Genis Laval

A l'aide d'un spectromètre à réseau tournant (2,5 Å de résolution) au foyer Cassegrain du télescope de 80 cm à l'Observatoire de Haute Provence, nous avons observé la Nova Cygni 1975 du 3 au 12 Septembre 1975, dans les régions spectrales λλ 3100-5000 et λλ 6200-8500 (Fig. 1). Les raies d'émission très élargies montrent toutes des centres d'absorption du côté des courtes longueurs d'onde, ce qui indiquerait la présence de plusieurs couches discrètes dans l'enveloppe. Le continuum de Balmer est parfaitement net, avec une forte absorption vers 3580 Å, analogue à celle observée dans Nova Her 1963 ou dans CH Cygni.

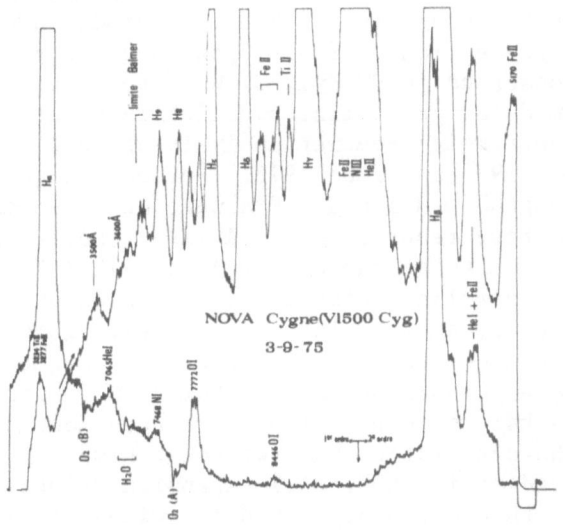

M. Friedjung (ed.), Novae and Related Stars, 177. *All Rights Reserved.*
Copyright © 1977 by D. Reidel Publishing Company, Dordrecht, Holland.

NOVA CYGNI 1975

J.B. Hutchings
Dominion Astrophysical Observatory

I have a few separate remarks and preliminary results to
report on Nova Cygni 1975, based on spectrographic observations
made in Victoria. 1) We have performed an analysis of line
profile changes similar to that made by Campbell (1976) for
several nights on which many spectrograms were obtained, between
Aug. 31 and Sept. 8. We find that analysis in terms of a simple
pulsating model with the $\sim$ 3 hour period does not yield consistent
results from night to night. Consequently we have investigated
a more complex model in which a cosine modulated beam rotates in
the centre of the nebula, to correspond to orbital motion of a
hot spot in a conventional C.V. model. While there are many free
parameters, it was found that such a model can reproduce all the
profile changes studied and implies a smaller nebula and later
"date of outburst" than Campbell's model. 2) I want to make a
point of noting the 2% change in the 3 hour period, which may
imply considerable precession of a hot spot, and the possibly
similar situation observed in VW Hyi. 3) We have attempted to
follow the 3 hour modulation of line profiles late into the
nebular stage, by performing Fourier analysis of spectrograms
obtained in November 1975 and June 1976. Periodicities in the
profiles of "wavelength" $\sim$ 3 A and 0.9 A were detected in the
data of these two dates, respectively. A line connecting these
frequencies has value zero at the end of Aug. 1975 ($\pm$ 10 days),
strongly suggesting that line emission is modulated by the
central variable source, at very large distances. This has
implications for the understanding of the nebula in general.
4) A study of the mean nebular profiles shows a more rapid decay
of the inner two peaks. A model which reproduces most of the
observed profile characteristics has two polar blobs ($\pm$ 20°) and
a ring ($\pm$ 10°) viewed at i $\simeq$ 60°. All expansion velocities are
about 700 km/sec, although the broad profile wings may imply a
velocity gradient in the equatorial regions.

M. Friedjung (ed.), Novae and Related Stars, 178. All Rights Reserved.
Copyright © 1977 by D. Reidel Publishing Company, Dordrecht, Holland.

MULTI-BAND PHOTOMETRY OF NOVA CYGNI 1975

K. KAWARA, T.MAIHARA, K.NOGUCHI, N.ODA, H.OKUDA,
S.SATO, M.OISHI*, and T.IIJIMA**
DEPARTMENT OF PHYSICS, KYOTO UNIVERSITY,KYOTO
*ASTRONOMICAL INSTITUTE, TOHOKU UNIVERSITY, SENDAI
**DEPARTMENT OF PHYSICS, NAGOYA UNIVERSITY,NAGOYA

Infrared and optical photometries of Nova Cygni 1975 were carried out from 1975 September 2, immediately after the light maximum, until November 10, the 69th day after the maximum.

Light curves of the nova are expressed by power functions of time with power indices, α -1.5.

The energy curves after the correction of interstellar extinction do not fit well blackbody radiation.

In the near infrared region, they show wavelength dependencies of λ^{-2}, due to the free-free emission from the circumstellar envelope of the nova, and in the optical region, they are influenced by emission lines.

This nova did not show any remarkable enhancements at 3.5 and 10.6 microns, indicating no substantial production of dust particles in the nova explosion.

M. Friedjung (ed.), Novae and Related Stars, 179. *All Rights Reserved.*

CLUES TO THE BINARY NATURE OF NOVA CYG 1975

Irene Kupo and Elia M. Leibowitz

Department of Physics and Astronomy
and the Wise Observatory
Tel-Aviv University, Tel-Aviv, Israel

Wavelength measurements of emission and absorption features in the spectrum of Nova Cyg 1975, observed in the Wise Observatory between August 31, 1975 and August 17, 1976, are interpreted on a basis of a binary model for the star. The structure of the emission lines lasting almost a year can be described as follows:

1. A broad component indicating an expansion velocity of 1900 km/s shifted to the violet by -90 km/s

2. Four sharp components superimposed on the diffuse line.

Average radial velocities of the outer and inner pairs of the sharp components have values of a few hundred km/s, about 1/10 of the expansion velocity. The two averages are different from each other as well as from the center of the broad component. The difference between the two mean velocities is in the range of the orbital velocity of a close binary star. We therefore suggest that the nova is a member of a binary system. The ejection of material from the nova took place in three distinct major events. One was a continuous outflow lasting at least one orbital period and forming a shell of spherical or axial symmetry with respect to the binary system. The broad emission component originates in this shell. Two other events were of a very short duration, of the order of an hour, during which pairs of clouds were ejected in diametrically opposite directions, in symmetry with respect to the ejecting star. With a few simplifying assumptions we estimate several parameters of the binary system. We discuss the possibility of generalizing the model for the explanation of observations in other novae, comparing it to other models of nova ejection that have been proposed.

M. Friedjung (ed.), Novae and Related Stars, 180. All Rights Reserved.

SPECTRAL EVOLUTION OF NOVA V 1500 CYG (1975) FROM SEPTEMBER 1975 TO MAY 1976

L.ROSINO

ASIAGO ASTROPHYSICAL OBSERVATORY OF THE UNIVERSITY OF PADOVA

Spectroscopic observations of Nova V 1500 Cyg have been carried out at Asiago after the first announcement of its discovery from August 30 to the present time.The photoelectric observations discussed by P.TEMPESTI prove : that the epoch of maximum was August 30.7 (JD 244 2655.2) with visual magnitude 1.7.The times employed by the nova to drop of 2 and 3 magnitudes were t_2= 2.5; t_3= 3.9.The spectrum near maximum and during the first decline was characterized by the presence of wide emission bands,shortward accompanied by broad diffuse absorption features with mean radial velocities increasing from -1700 to -4200 km/s on Sep 23 when the absorption disappeared.The emission bands were very broad and many of them overlapped.The expansion velocity derived by the halfwidth of the bright bands increased from 860 to 1500 km/s on Sep 3,then remaining sensibly constant.The emission bands were saddle-shaped with two symmetric peaks at each side of the central minimum.At first,the strongest bands in the spectrum were due to H,CaII (H and K),NaI,FeII.In the infrared the Paschen series , OI-NI and particularly OI 8446,were outstanding.The nova entered in the nebular stage between Sep 9 and 12.This phase was characterized by the strengthening of the forbidden lines of NeIII 3869-3968, OIII 4363, 4958-5007, NII 5754 and permitted lines of NIII 4640,HeI and HeII 4686,which soon became prominent.Later,at the end of Sep and beginning of Oct,also forbidden lines of NeV, CaV,FeVI and FeVII appeared and rapidly strengthened.At the end of the year the forbidden line of FeVII at 6087 was still rising, being already as strong as HeI 5875,but it went slowly declining from January to May 1976.On May 15 the lines 6678 and 7065 of HeI were barely visible,while 5875 was weaker than FeVII 6087.In the infrared spectrum the lines of OI,NI and even 8446 have disappeared,while the forbidden line OII 7317-30 was outstanding.

M. Friedjung (ed.), Novae and Related Stars, 181. *All Rights Reserved.*
Copyright © 1977 *by D. Reidel Publishing Company, Dordrecht, Holland.*

MODEL OF A NOVA OUTBURST ON THE BASIS OF MEASUREMENTS ON V1500 CYGNI

Waltraut Carola Seitter

Astronomisches Institut der Universitaet Muenster, FRG

Spectroscopic and photometric observations of Nova Cygni 1975 yield a temperature curve and a luminosity curve which in turn permit the derivation of the growth of radius of the nova quasi-photosphere. Over a period of about two days the photospheric radius increases from less than 1 $R_\odot$ to almost 500 $R_\odot$. Most interesting is the fact that the rate of growth corresponds to a constant acceleration rather than a constant velocity which would be expected if the expansion occurred with the particle velocity of 1500 km sec^{-1}, a value that stayed practically constant over the above-mentioned time interval.

Constant acceleration points to a constant driving force. It might be identified with constant radiation pressure from the surface of the central star after material has been lifted off and expands as a thin nebula. Within this pre-ejected nebula the photosphere appears as a travelling condensation front which eventually catches up with the outer boundary of the nebula. This is the time of maximum.

The above model also explains the higher temperature and more extended atmosphere (higher luminosity class) during the earlier pre-maximum stages.

M. Friedjung (ed.), Novae and Related Stars, 182. All Rights Reserved.

PRE-MAXIMUM SPECTRA OF V1500 CYGNI

Waltraut Carola Seitter

Astronomisches Institut der Universitaet Muenster, FRG

Spectra of Nova Cygni, beginning with UT August 29.83,1975, display nova characteristics of an earlier state of evolution than hitherto observed for any nova. While higher dispersion plates show few absorptions, I-N plates of dispersion 240 Å/mm reveal a wealth of broad absorption features, in particular, N II and O II blends of unusual strengths, matching almost those of the hydrogen lines. Weak H-emission is indicated. The probable absence of higher ionization stages fixes the temperature around B0 at the beginning of the first night. At the end of the night it appears to be near B3. Abundance effects cannot account for the abnormal line strengths since these disappear within the first nine hours of observation, a time scale too short to suggest noticeable abundance variations. The spectral peculiarities must then be attributed to luminosity effects. Tentative extrapolations of line strengths in normal stars lead to luminosity classifications on a scale extending considerably beyond class Ia$^+$. The same process is applied to spectra of the second night when the nova resembled a medium A-type star of high luminosity.

The spectral history of the nova before maximum permits the following conclusions:
1) The observed spectral type is earlier the earlier the evolutionary state at the time of observation.
2) The luminosity class is higher the earlier the time of observation.
3) Extrapolation of the observed temperature curve leads to a prenova temperature of about 40 000°K.

In a subsequent paper the above relations are used to derive a new model of nova outbursts.

M. Friedjung (ed.), Novae and Related Stars, 183. *All Rights Reserved.*
Copyright © 1977 by D. Reidel Publishing Company, Dordrecht, Holland.

Ultraviolet Photometry of Nova Cygni 1975

Chi-Chao Wu

Kapteyn Astronomical Institute, Dept. of Space Research,
University of Groningen, P.O. Box 800, Groningen, The
Netherlands

Nova Cygni was observed 28 times at 100 days after visual maximum
with the ultraviolet experiment on board the Astronomical Netherlands Satellite. The instrument is a 5-channel intermediate band
(100-200 A) spectrophotometer. Central wavelengths for the
channels are 1550, 1800, 2200, 2500 and 3300 A.

Fourier analysis of the data points yields a period of 0.140
days. The ultraviolet and ground based light curves suggest a
geometric model similar to that the for dwarf nova U Gem. It can
be shown that at the 1800, 2200 and 2500 A channels, the intrinsic
spectral energy distribution of the nova (remnant + shell) is
smooth. Since the interstellar extinction is peaked at 2200 A,
there is a strong deficiency of flux at this channel. The observed
spectrum is dereddened until a smooth flux distribution is obtained between 1800 and 2500 A. $E(B-V) = 0.69 \pm 0.03$ is derived.
Interstellar absorption lines give similar amount of reddening,
indicating that there is no appreciable dust formation in Nova
Cyg. A temperature of 65000 K is also given by the dereddened
spectrum. The average distance obtained by various methods is
1550 pc. With these parameters, the luminosities at maximum and
100 days after are 500,000 and 30,000 solar luminosities, respectively. So unlike FH Ser, constant luminosity was not maintained for the first 100 days. However, during this period, the
visual brightness decreased by a factor of 6600 while the bolometric luminosity decreased only by a factor of 20; and the
effective temperature increased from 10000 to 65000 K. So the
optical decline is mostly due to the shift of the flux maximum
into the ultraviolet.

NOVA CYGNI 1975 AS A CLASSICAL THERMONUCLEAR RUNAWAY EVENT

Sumner Starrfield
Arizona State University

Warren M. Sparks
Goddard Space Flight Center

James W. Truran
University of Illinois

J. S. Gallagher
University of Minnesota

P. A. Strittmatter
University of Arizona

Hugh M. Van Horn
University of Rochester

We discuss the various features of the optical and spectral developments of Nova Cygni 1975 (V1500 Cygni). The range of more than 19 magnitudes, the rounded maximum and steep decline and the absence of post-maximum absorption systems establish this as a unique object. We are able to explain these features if we assume that a thermonuclear runaway has taken place in the carbon-enhanced, hydrogen-rich, envelope of a single white dwarf where the hydrogen has been accreted by passage of the white dwarf through an interstellar cloud. This would be possible for Nova Cyg since it lies directly in the plane of the galaxy. The carbon enhancement, necessary for ejection, has occurred through the action of a surface convection zone which has mixed a fraction of the accreted hydrogen into the edge of the carbon core. We shall also discuss the observed oscillations and present alternative models for the object.

M. Friedjung (ed.), Novae and Related Stars, 185. All Rights Reserved.
Copyright © 1977 by D. Reidel Publishing Company, Dordrecht, Holland.

CONCLUSIONS OF SESSION AND SUMMARY OF DISCUSSION

"OBSERVATIONS OF NOVA CYGNI 1975"

M. Friedjung

Institut d'Astrophysique
98 bis, Bd Arago
75014 Paris, France

The discussion was centred on two problems. The first was which model was most appropriate for this unusual nova. The very rapid rate of fading and claimed absence of diffuse enhanced and Orion absorption might suggest nearly instantaneous ejection, the large absorption line widths indicating instantaneous ejection type II. The increase of absorption velocity with time could then be explained for absorption mainly produced in regions optically thick to Lyman continuum radiation, if the envelope became more ionized with time, so only high velocity gas near the outer edge would absorb in later stages. However Duerbeck stated that a diffuse enhanced spectrum was clearly present on Sept. 1.1 (ESO spectra), while reports of the detection of Orion lines in Japan were also mentioned. The remnant detected by Wu in the UV 3 months after maximum had a radius of 1×10^{11}cm, too large to be a white dwarf or conventional hot spot (Smak), and perhaps was a photosphere formed by continued ejection (Bath), an accretion disc, or a bloated stellar remnant.

The other problem discussed was the nature of the 0.14 day variations. Hutching's talk showed that the Hα profile changes found by Campbell need not contradict an eclipse explanation of the continuum variation. Smak suggested that they were produced by an extended envelope, asymmetric because of the binary nature of the nova, or were due to heating of part of the secondary component, whose visibility would vary.

Other points raised included the disagreements of E(B-V) determinations. The method of Neff who compared the observed Sept. 1975 continua with theoretical nebular continua, may be less valid than the UV one of Wu. The single star interpretation of Nova Cygni also encountered opposition.

M. Friedjung (ed.), Novae and Related Stars, 186. *All Rights Reserved.*
Copyright © 1977 by D. Reidel Publishing Company, Dordrecht, Holland.

PART V

THEORIES OF THE CAUSES OF OUTBURSTS

A REVIEW OF THE THERMONUCLEAR RUNAWAY MODEL OF A NOVA OUTBURST

Warren M. Sparks
Goddard Space Flight Center

Sumner Starrfield
Arizona State University

James W. Truran
University of Illinois

The diagram shown in Figure 1 is a familiar one to all students as the nova phenomena. It illustrates the widely accepted

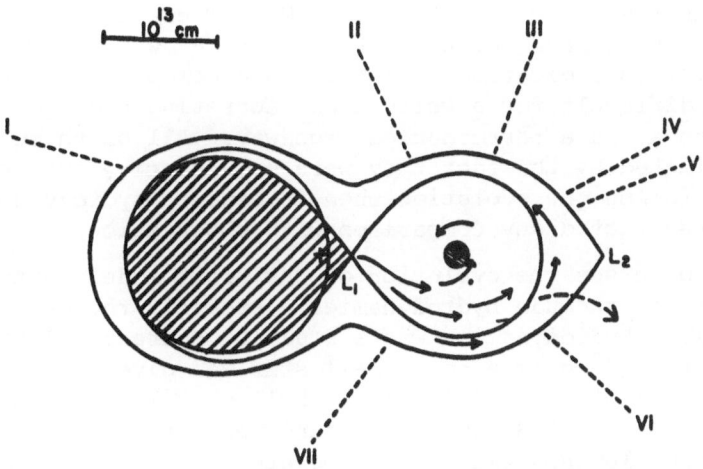

Fig. 1 - Schematic diagram of Kraft's (1963) binary nova model.

M. Friedjung (ed.), Novae and Related Stars, 189-204. All Rights Reserved.
Copyright © 1977 by D. Reidel Publishing Company, Dordrecht, Holland.

model put forth by Kraft (1963). During this conference we have heard talks by D. Lin and by G. Bath on how material can flow off the red companion and form a ring around the white dwarf. We have also heard how material in the ring spirals inward, giving up its potential energy, and eventually accrets onto the white dwarf. In fact, this potential energy has been proposed by Bath (1976) as the source of energy for the dwarf nova outburst.

I will limit my talk to what happens to this hydrogen-rich material after it has settled onto the white dwarf. Kraft (1963) proposed that this material eventually becomes degenerate. This degenerate hydrogen-rich matter heats up as more material is accreted on top of it and eventually reaches ignition temperatures. Now degenerate material has the property that to the first order the pressure is not a function of the temperature. Thus, the energy generation is not limited by expansion until degeneracy is lifted. This leads to a thermonuclear runaway which was proposed by Kraft as the cause of the nova outburst.

There were a large number of hydrostatic studies which confirmed this general behavior. Among these are Giannone and Weigert (1967), Rose (1968), Saslaw (1968), Starrfield (1971a, 1971b), Redkoborodyi (1972), Taam and Faulkner (1975), and Colvin, Van Horn, Starrfield and Truran (1976). The basic conclusion of all of these studies is that a thermonuclear runaway will occur under these conditions and that the energy released is comparable to that of a nova outburst. The envelope mass required lies in the range of 10^{-4} to $10^{-3} M_\odot$. These studies included a variety of mass accretion rates, white dwarf masses and other parameters indicating that this is a common occurrence. In fact, it is difficult for a white dwarf accreting hydrogen-rich material to avoid a thermonuclear runaway. All of these studies were limited by the fact they were hydrostatic, and thus they could not follow the evolution when it becomes hydrodynamic. This severely restricted any comparison with observations.

In order to pursue the evolution of the Kraft model further, it was necessary to include hydrodynamics. Probably the earliest attempt to include hydrodynamics was a study by Pottasch (1959) in which the outer 10^{29}g of a hot, small star was given an initial velocity of 800 km sec^{-1}. This shell was also allowed to expand with a velocity of 3 times the sound velocity. The calculated surface luminosities and temperatures compared favorably with observations. In a different approach Nadezhin and Frank-Kamenetskii (1963) forced a collapse in an equilibrium polytropic model which resulted in an outward-propagating shock wave. This shock wave imparted a velocity greater than the escape velocity to an outer fraction of the star. The mass, kinetic energy, and average velocity of the ejected material were calculated and then compared with observations in order to find the total mass of the star.

The research for my doctoral thesis (Sparks 1969) consisted
of putting an excess amount of energy into a stellar envelope and
evolving it with an explicit hydrodynamics code. Figure 2 shows

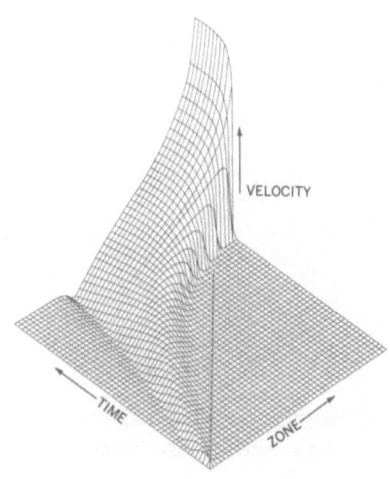

Fig. 2 - Projection of a three-dimensional plot of material
         velocity vs. time and zone number for shock ejection.

what happens when this energy is put into a region in a time
interval that is shorter than the sound travel time across that
region. It is a projection of a three-dimensional plot of
material velocity as a function of time and zone number (radius).
A shock wave is formed at the bottom of the envelope and propagates outward. The curving to the right shows the acceleration
of the shock wave and the increase in the material velocity
indicates the increase in the shock strength as it propagates
through the less dense outer zones. After the passage of the
shock wave the ejected material is left with a steep velocity
gradient (cf. Hazelhurst 1962). If the energy is deposited in a
region in a time interval that is longer than the sound travel
time across that region, then a "pressure" wave develops.

Figure 3 shows how the "pressure" wave lifts the outer zones
and gives them roughly the same velocity.   In Figure 4, a plot of

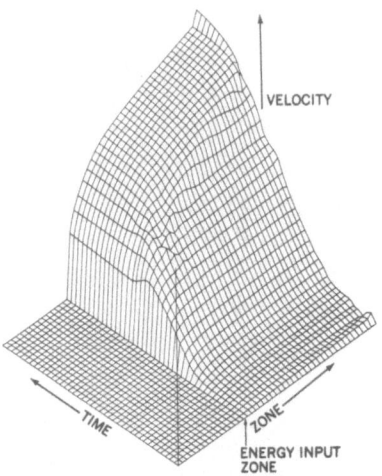

Fig. 3 - Projection of a three-dimensional plot of material
         velocity vs. time and zone number for pressure ejection.

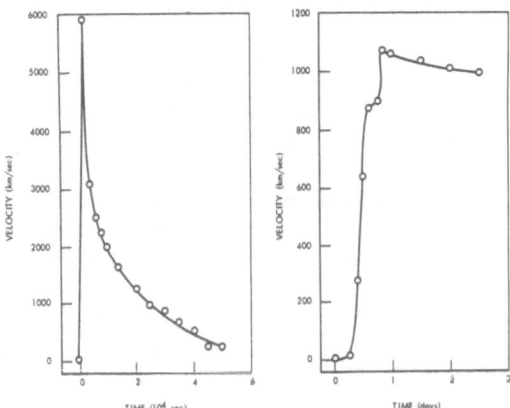

Fig. 4 - Velocity of the material at an optical depth of 2/3 in
         the continuum vs. time. Shock ejection is on the left;
         pressure ejection, on the right.

the material velocity at an optical depth of 2/3 in the continuum
as a function of time is given. On the left side shock ejection
causes a very rapid rise, and then the velocity decreases as the
effective radius moves deeper into the material, where the velocity
is slower, when the density decreases due to expansion. The
pressure-ejected material on the right side shows a more constant
velocity after its increase because the material is expanding
with nearly the same velocity. We cannot compare these velocities
directly with observed velocities since they are derived from
spectral lines. However, we would expect the general shapes of
the velocity curves to be similar. In Figure 5, the light and

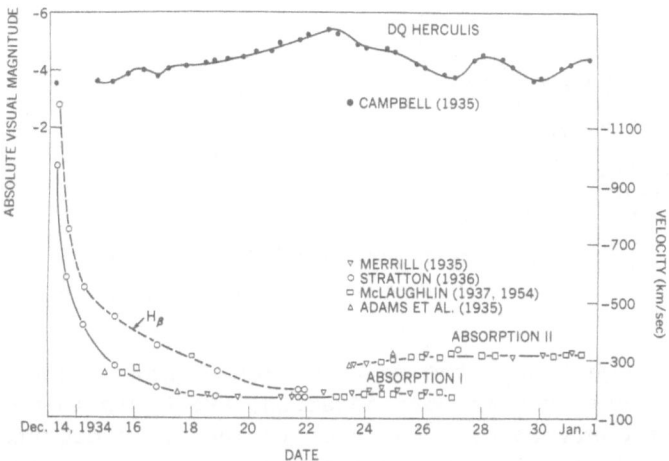

Fig. 5 - Light curve and velocity curve of DQ Her. Absorption
         I and II are the pre-maximum and the principal spectra
         respectively. The dashed line is the velocity of Hβ.

velocity curves of DQ Her are presented. The pre-maximum
absorption system system (Absorption I) is suggestive of shock-
ejected material while the principal absorption system (Absorption
II) is similar to pressure-ejected material.

     Rose and Smith (1972) also studied shock ejection with an
explicit hydrodynamics computer code. In their study excess
energy was injected into a stellar envelope by multiplying the
equilibrium CN cycle rate by a numerical factor that increased
with time. They concluded that shock ejection is probably not the

sole cause of mass loss in novae because it requires very high
rates of nuclear energy generation and gives the wrong velocity
distribution.   Then they suggested that the outburst is due to
radial pulsations but were unable to produce a model that ejected
any material.

I now wish to discuss the research that Sumner Starrfield,
James Truran and I have done on Kraft's thermonuclear runaway
model (see Starrfield, Sparks and Truran, 1976 and references
therein).   Although several other groups are now doing similar
studies, as we shall hear about later today, their work has not
been published yet, and I am not familiar with it.   It became
obvious that a hydrodynamics computer code with the proper energy
generation was necessary to study this model.   The code that we
used was an implicit hydrodynamics Lagrangian computer code
developed by Sigfried Kutter and myself (1972).   The equations
solved simultaneously are:

$$\frac{4\pi}{3} \frac{\partial r^3}{\partial m} = V , \qquad \text{conservation of mass ;} \quad (1)$$

$$\frac{\partial u}{\partial t} + 4\pi r^2 \frac{\partial}{\partial m} (P + q) = -\frac{Gm}{r^2} , \quad \text{conservation of momentum ;} \quad (2)$$

$$\frac{\partial E}{\partial t} = \epsilon - \frac{\partial L}{\partial m} - (P + q) \frac{\partial V}{\partial t} , \quad \text{conservation of energy ;} \quad (3)$$

$$L = -\frac{256\sigma\pi^2}{3} \frac{r^4 T^3}{\kappa} \frac{\partial T}{\partial m} + \pi \left(\frac{G}{2}\right)^{1/2} k^{-3/2} l^2 C_P \left(\frac{\partial \ln V}{\partial \ln T}\bigg|_P\right)^{1/2} \frac{rT m^{1/2}}{V} (\nabla - \nabla_{ad})^{3/2} ,$$

$$\text{energy transport by radiation and convection ;} \quad (4)$$

$$\partial r / \partial t = u , \qquad \text{definition of velocity.} \quad (5)$$

The first four equations, the conservation of mass, momentum and
energy and the energy transport equation, are basic in most
stellar evolutionary codes.   In order to include hydrodynamics
one needs the addition of the definition of velocity (equation 5),
the acceleration term in the conservation of momentum equation and
the artificial viscosity pressure term, q, to handle shock waves.

Figure 6 shows our nuclear reaction network for the energy
generation.   Under ordinary stellar energy generation conditions
the β^+-decay rates are much more rapid than the proton-capture
rates such that the possibility of ^{13}N capturing a proton need
not be considered.   In fact the abundances of ^{13}N, ^{14}O, ^{15}O
and ^{17}F are usually not even calculated.   For the nova case
we have found that the temperature become so high that the proton-
capture rates are faster than the β^+-decay rates.   Thus, the
abundances of ^{13}N, ^{14}O, ^{15}O and ^{17}F must be taken into account.
This is called the fast CNO cycle on which J. Audouze will talk
later today.   When the proton-capture rates become very short, the
energy generation rate is limited by the β^+-decay rates.   For a
solar abundance of CNO the equilibrium energy generation rate is
limited to 1.8×10^{14} erg g^{-1} sec^{-1} no matter how high the temperature and density are.   We (Starrfield, Sparks, and Truran 1974)

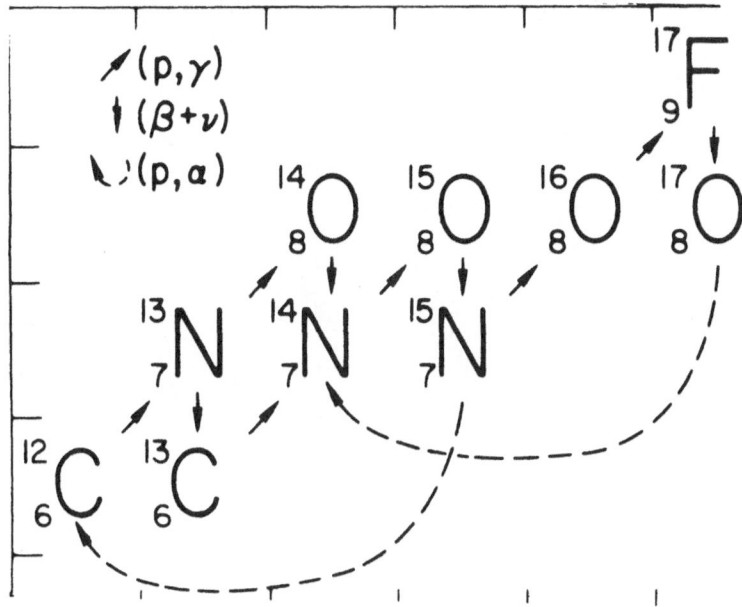

Fig. 6 - The nuclei and reactions included in the network which
was used to obtain the energy generation rates and follow
the changes in abundance.

have shown that for a burst type of ejection the energy generation
rate at maximum required is approximately equal to the gravity
of the white dwarf times the velocity of ejection, which is
typically 10^{16} erg g^{-1} sec^{-1}. A burst type of ejection means that
the maximum energy generation rate is much greater than the
luminosity gradient ($\varepsilon_{nuc} >> \frac{\partial L}{\partial m}$) but does <u>not</u> imply shock ejection.

It appears that fast novae require this type of ejection, although
the condition on the energy generation rate may not be this
severe since all of the material is not ejected. Thus, in order
to obtain a large energy generation rate at maximum, the number
of CNO nuclei must be increased. I will return to the point of
CNO enhancement later in this talk.

The initial model which we have used is a white dwarf with
a 10^{-4} to $10^{-3}M_{\odot}$ hydrogen-rich envelope in hydrostatic and
thermal equilibrium. In the model that I will discuss the
bottom zone of the hydrogen envelope contains 15% carbon and 15%
oxygen. If this is mixed throughout the envelope it amounts to a

CNO enhancement of approximately 10 times solar.   Figure 7 is a

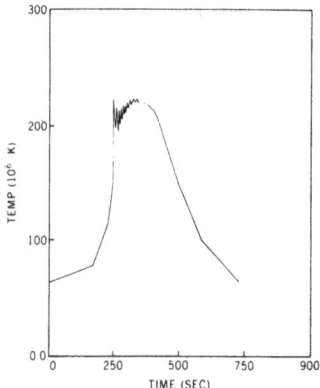

Fig. 7 - The temperature of the bottom hydrogen zone as a
         function of time.

graph of the temperature of the bottom zone of the hydrogen-rich
envelope as a function of time for this model. This temperature
in the equilibrium model actually starts at about 10 million
degrees taking a few thousand years to reach 30 million degrees,
only a couple of more days to reach 100 million degrees and
then only seconds to reach 200 million degrees. Clearly the
model is undergoing a thermonuclear runaway. This graph demon-
strates the necessity of an implicit hydrodynamics computer code
to study this phenomena. Implicitness is required for the very
long time steps necessary for the evolution to reach thermonuclear
runaway, and, with this very rapid rise in temperature, obviously
hydrodynamics become important. The rapid oscillation in the temperature is due to motions in the envelope as it expands. This
expa..sion eventually causes the cooling of this zone.

     Figures 8 and 9 show the isotopic abundances as a function of
time for the bottom zone of the hydrogen-rich envelope. The
first point plotted for each nucleus represents the initial
abundance of that nucleus at the beginning of the evolutionary
sequence, not its abundance at zero seconds on the figure.
Nitrogen -13 first increases as ^{12}C proton captures and then
later decreases as it starts to proton capture to ^{14}O. The rapid
oscillations in various nuclei are due to the oscillation in the

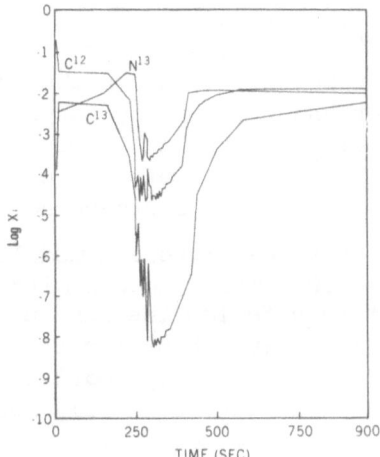

Fig. 8 - The abundances by weight of ^{12}C, ^{13}C and ^{13}N as functions of time for the bottom hydrogen zone. The initial point for each nucleus marks its abundance at the beginning of the evolution, not at time zero on the graph.

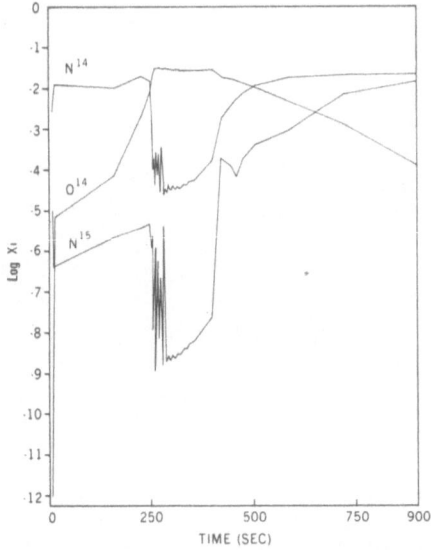

Fig. 9 - The abundances by weight of ^{14}N, ^{14}O, and ^{15}N as functions of time for the bottom hydrogen zone. The initial point for each nucleus marks its abundance at the beginning of the evolution, not at time zero on the graph.

oscillation in the temperature shown in Figure 7. Near maximum
energy generation most of the CNO nuclei are ^{14}O and ^{15}O. As the
envelope expands and cools, they decay to ^{14}N and ^{15}N, respectively. This leads to a $^{14}N/^{15}N$ ratio which is less than 1 in the
inner most zone! These nuclei are mixed to the outer layers of
the envelope by convection and their decay supplies the final
energy which ejects most of the material. Depending upon the
energetics of the model and the mass of the envelope, the $^{14}N/^{15}N$
ratio in the ejecta may reach values approaching 1.

The early evolution of three different models on the H-R
diagram is shown in Figure 10. These models differ in the initial
intrinsic luminosity, but they follow the same evolutionary path.
In fact, they retrace their evolution up their cooling curves.
However, the lowest intrinsic luminosity model takes a much longer
time to reach thermonuclear runaway, basically because of its
lower temperatures in the energy generating region (see Truran,
Starrfield, Strittmatter, Wyatt, and Sparks 1976, for further
detail). Figure 11 shows the time to thermonuclear runaway as
a function of intrinsic luminosity of the white dwarf. If the
time between runaways is on the order of thousands of years then
the intrinsic luminosity of the white dwarf must be low.

The light curve of one of our models is shown in Figure 12.
The initial rise is due to a shock wave caused by the thermonuclear

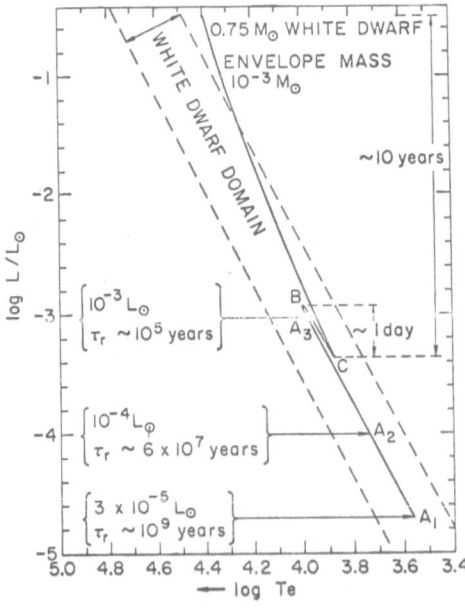

Fig. 10 - Evolution in the HR diagram of a $0.75M_\odot$ white dwarf of
          hydrogen envelope mass $10^{-3}M_\odot$ as a consequence of ther-
          monuclear runaway for three different initial intrinsic
          luminosities.

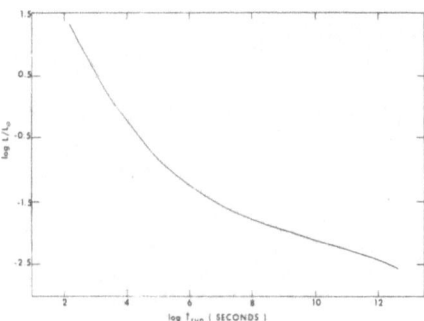

Fig. 11 - The time for a white dwarf to evolve from the equilibrium model to a temperature in the shell source of 30 million degrees as a function of its initial intrinsic luminosity.

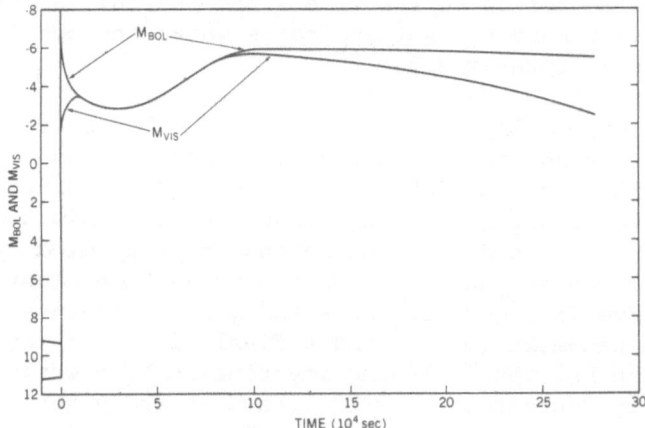

Fig. 12 - The computed bolometric and visual light curves of a hydrodynamic nova model.

runaway. This shock wave does not eject any material. The luminosity remains high after the shock wave because of heating from the decay of the β^+-unstable nuclei. The effect of this heating is demonstrated by the positive luminosity gradient in the outer layers shown at a time of 2×10^3 sec in Figure 13. (All times refer to the time after the initial rise.) At a later time, 1.9×10^4 sec, all of the β^+-unstable nuclei have decayed

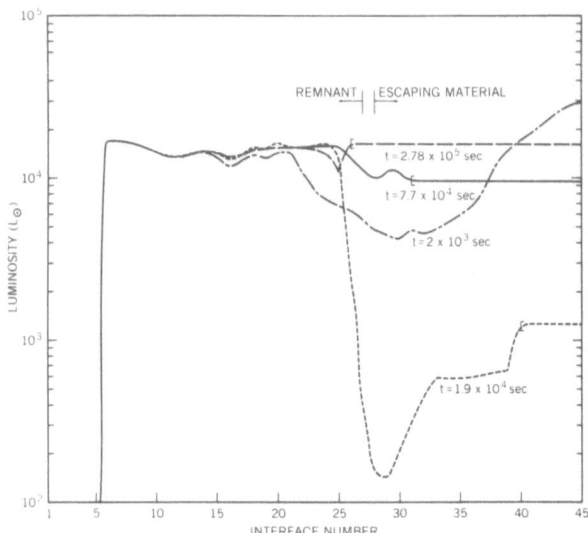

Fig. 13 - Luminosity as a function of interface number and time.
           The time refers to the time since the initial rise and
           the left-hand bracket indicates where the continuum
           optical depth is 2/3.

and the outer escaping layers have cooled due to expansion and radiative loses producing the dip in the light curve shown in Figure
12. By 2.78×10^5 sec the high inner luminosity has progressed
outward giving the final rise in the light curve. This high
inner luminosity is caused by a non-degenerate rekindled hydrogen-
burning shell source at the bottom of the remnant envelope. The
visual light curve in Figure 12 shows the general features of an
initial rise, a pre-maximum halt and a final rise of a fast nova.
After maximum the bolometric luminosity remains high while the
visual luminosity decreases. This decrease in the visual luminosity is best explained by considering Figure 14 which is a plot
of the radius as a function of interface number. From 7.7×10^4 sec
to 2.78×10^5 sec when the bolometric luminosity is constant the
photospheric radius (where $\tau = 2/3$ and shown by the left-hand
bracket) is decreasing because the ejected material is becoming
optically thin. This causes the effective temperature to
increase and less radiation falls in the visual region. In order
to follow the evolution of the remnant we removed the escaping
material from the calculations and evolved only the remnant
envelope. This soon reached the "equilibrium model" shown in
Figure 14 with a luminosity of $1.7 \times 10^4 L_\odot$. This compares favorably with the constant bolometric luminosity of $1.6 \times 10^4 L_\odot$ found
by Gallagher and Code (1974) from ultraviolet observations of FH Ser

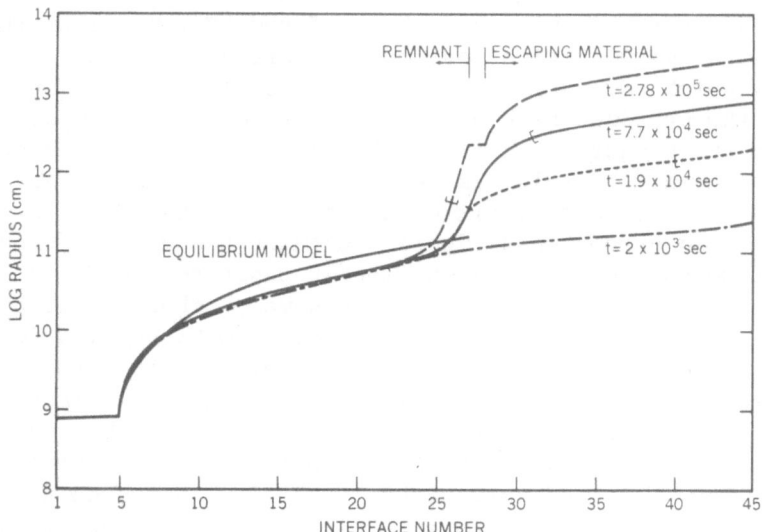

Fig. 14 - Radius as a function of interface number and time.
          The time refers to the time since the initial rise and
          the left-hand bracket indicates where the continuum
          optical depth is 2/3. The "equilibrium model" is the
          remnant after it has nearly reached hydrostatic and
          thermal equilibrium.

from 4.4 to 57 days after maximum. This "equilibrium model" also
has an effective temperature of 63000°K which agrees well with
the value of 65000°K found by Wu and Kester (1976) for Nova Cygni
100 days after visual maximum. They also concluded that for
Nova Cygni the decrease in visual light is mostly due to the
shift of flux maximum to the ultraviolet.

    We find that roughly 10% of the $10^{-3} M_\odot$ envelopes (Starrfield,
Sparks, and Truran 1974) and 70% of the $10^{-4} M_\odot$ envelopes
(Starrfield, Sparks, and Truran, 1977) are ejected with velocities
of 200-2000km sec^{-1} which also agree with observations. As
previously stated, the theoretical luminosities and effective
temperatures of the remnant envelopes closely match observations
up to a hundred days after maximum. However, the lower observed
bolometric luminosities of two 50-year old novae (Gallagher and
Holm 1974) indicates that the remnant envelope does not evolve
on a nuclear time scale. Thus the remnant envelope is probably
expelled. This can be accomplished by either gravitational
stirring by the binary system within the remnant envelope (Sparks,
Starrfield, and Truran 1976a) or by radiation pressure from the

rekindled hydrogen-burning shell source (Sparks, Starrfield, and Truran 1976b).

Let us return to the enhancement of the CNO nuclei at the bottom of the hydrogen-rich envelope. There are a number of methods by which CNO nuclei from the core could be mixed into the bottom of the hydrogen-rich envelope. Durisen (1976) suggested that shear turbulence could mix in CNO nuclei into the accreting hydrogen material. Colvin, Van Horn, Starrfield and Truran (1976) have found that convective mixing during the accretion of hydrogen-rich gas onto a carbon white dwarf increases the carbon abundance in the bottom of the accreted envelope but not to the extent necessary for our fast nova models. However, other models are being investigated. Finally, the convective overshooting during the thermonuclear runaway can bring CNO nuclei up from the core. Although there has been some criticism of our models because we require a CNO enhancement, it is now clear that the observations give strong support for such a CNO enhancement. For example, Pottasch (1959) observed that carbon was solar, nitrogen forty-five times solar, and oxygen five times solar for five novae, while Mustel' and Boyarchuk (1959) and Mustel' and Baranova (1965) found that the CNO nuclei were approximately one hundred times solar for DQ Her. For the recent slow nova HR Del 1967, Ruusalepp and Luud (1971) found that carbon and nitrogen were ten times solar, Sanyal and Robbins (1975) found that oxygen was greater than three times solar, Gallagher and Anderson (1976) found that nitrogen was two or more times solar, and Antipova (1974) found that carbon was ten times solar while nitrogen and oxygen were one hundred times solar. R. Tylenda and S. Collin-Souffrin have both reported earlier in this conference large overabundances in nitrogen and oxygen for HR Del during the nebular stage. The range in the values determined for HR Del indicate the difficulties associated with such observations. Nevertheless, we feel that the large number of observations of CNO overabundances in novae provide strong support for our theoretical models.

Another of our theoretical predictions is that the concentrations of ^{13}C, ^{15}N and ^{17}O will be greatly enhanced in nova ejecta. Recently, Sneden and Lambert (1975) have set upper limits on the $^{13}C/^{12}C$ and $^{15}N/^{14}N$ ratios which they claim are in conflict with our results. The fact that strongly enhanced ^{15}N and/or ^{13}C are required to fit their line profiles of CN molecular bands is probably the strongest observational evidence of a thermonuclear runaway in a nova. This is in spite of the fact that the measurements were taken near maximum light, a time when it is difficult to determine exactly how much material is mixed up from the energy generation region and when the ejected material is mixing with the ring of unprocessed material surrounding the white dwarf. Thus it should not be surprising that their observations indicate lower ratios of $^{15}N/^{14}N$ and $^{13}C/^{12}C$ than our results which are averaged over the entire ejecta. Further discussions of this extremely interesting subject are found in Sparks, Starrfield, and Truran (1976a) and Starrfield, Truran and Sparks (1976).

REFERENCES

Adams, W. S., Christie, W. H., Joy, A. H., Sanford, R. F., and Wilson, O. C. 1935, Pub. A. S. P., 47, 205.

Antipova, L. I. 1974, Highlights in Astronomy, 3, 501.

Bath, G. T. 1976, I.A.U. Symposium No. 73, Structure and Evolution of Close Binaries, ed. P. Eggleton, S. Mitton, and J. Whelan (Reidel: Dordrecht).

Campbell, L. 1935, Harvard Bull., No. 898.

Colvin, J. D., Van Horn, H. M., Starrfield, S., and Truran, J. W. 1976, preprint.

Durisen, R. H. 1976, preprint.

Gallagher, J. S., and Anderson, C. M. 1976, Ap. J., 203, 625.

Gallagher, J. S., and Code, A. D. 1974, Ap. J., 189, 303.

Gallagher, J. S., and Holm, A. V. 1974, Ap. J. (Letters), 189 L123.

Giannone, P., and Weigert, A. 1967, Zs. f. Ap., 67, 41.

Hazelhurst, J. 1962, Adv. Astr. and Ap., 1, 1.

Kraft, R. P. 1963, Adv. Astr. and Ap., 2, 43.

Kutter, G. S., and Sparks, W. M. 1972, Ap. J., 175, 407.

McLaughlin, D. B. 1937, Pub. Michigan Obs., 6, 107.

_____. 1954, Ap. J., 119, 124.

Merrill, P. W. 1935, Ap. J., 82, 413.

Mustel', E. R., and Baranova, L. I. 1965, Sov. Astr.-AJ, 9, 31.

Mustel', E. R., and Boyarchuk, M. E. 1959, Sov. Astr.-AJ, 3, 744.

Nadezhin, D. K., and Frank-Kamenetskii, D. A. 1963, Sov. Astr.-AJ, 6, 779.

Pottasch, S. 1959, Ann. d'ap., 22, 310, 412.

Redkoborodyi, Y. N. 1972, Astrofizika, 8, 261.

Rose, W. K. 1968, Ap. J., 152, 245.

Rose, W. K., and Smith, R. L. 1972, Ap. J., 172, 699.

Ruusalepp, M., and Luud, L. 1971, Tartu Obs. Publ., 39, 89.

Saslaw, W. C. 1968, M.N.R.A.S., 138, 337.

Sanyal, A., and Robbins, R. R. 1975, private communication.

Sneden, C., and Lambert, D. L. 1975, M.N.R.A.S., 170, 533.

Sparks, W. M. 1969, Ap. J., 156, 569.

Sparks, W. M., Starrfield, S., and Truran, J. W. 1976a, Ap. J., 208, 819.

_____. 1976b, Bul. A. A. S., 8, 321.

Starrfield, S. 1971a, M.N.R.A.S., 152, 307.

_____. 1971b, ibid., 155, 129.

Starrfield, S., Sparks, W. M., and Truran, J. W. 1974, Ap. J. Suppl., 28, 247.

_____. 1976, I. A. U. Symposium No. 73, Structure and Evolution of Close Binaries, ed. P. Eggleton, S. Mitton, and J. Whelan (Reidel: Dordrecht).

_____. 1977, in preparation.

Starrfield, S., Truran, J. W., and Sparks, W. M. 1976, a paper presented at the CNO Conference at Grenoble.

Stratton, F. J. M. 1936, Pub. Solar Phys. Obs., Cambridge Univ. Vol. 4, part 4.

Taam, R., and Faulkner, J. 1975, Ap. J., 198, 435.

Truran, J. W., Starrfield, S., Strittmatter, P. A., Wyatt, S. P., and Sparks, W. M. 1976, to be published in the Ap. J.

Wu, C.-C., and Kester, D. 1976, preprint.

NUCLEOSYNTHESIS INDUCED DURING NOVA OUTBURST

Jean AUDOUZE[1,2] and Bernard LAZAREFF[2]

[1] Laboratoire René Bernas du Centre de Spectrométrie Nucléaire et de Spectrométrie de Masse, B.P. n° 1, 91406 ORSAY, France.

[2] Radio Astronomie, Observatoire de Meudon, 92190 MEUDON, France.

I - INTRODUCTION

As it has been shown in the previous communication (Sparks et al, this conference), the nova outburst is triggered by explosive CNO burning occuring in the outer regions of the prenova at temperatures $T > 10^8$ K and densities $\rho \sim 10^3 - 10^5$ g cm^{-3}. At such temperatures, nuclear reactions are faster than beta decays. The general properties of this " hot " CNO cycle are recalled in section II. In section III some specific calculations of nucleosynthesis based on nova models are presented and discussed. In some extreme nova models, temperatures as large as $T \sim 2 \cdot 10^9$ K and densities $\rho \sim 10^5$ g cm^{-3} might be reached. Section IV shortly reviews the possibility of neutron production suggested by Clayton and Hoyle (1974) to induce s and r element nucleosynthesis and the possible occurence of p process explosive nucleosynthesis. Finally in section V the possible role of novae in the chemical evolution of the Galaxy is discussed after the recent determinations of the H^{12}C^{15}N/H^{13}C^{14}N ratio in various interstellar clouds.

II - HOT CNO-Ne CYCLE

The hot CNO-Ne cycle designates the set of nuclear reactions occuring between hydrogen, helium and the $12 < A < 25$ nuclei at temperatures above 10^8 K. When the temperature rises up to $2 \cdot 10^8$ K, particle induced reactions become faster than beta decays thus change the location of bottlenecks (created by the slowest reactions in the CNO tri-cycle and the Ne-Na cycle (fig. 1). Hot CNO cycle favors the production of elements with beta unstable progenitors such as ^{13}C, ^{14}N, ^{15}N and ^{17}O.

M. Friedjung (ed.), Novae and Related Stars, 205-215. All Rights Reserved.

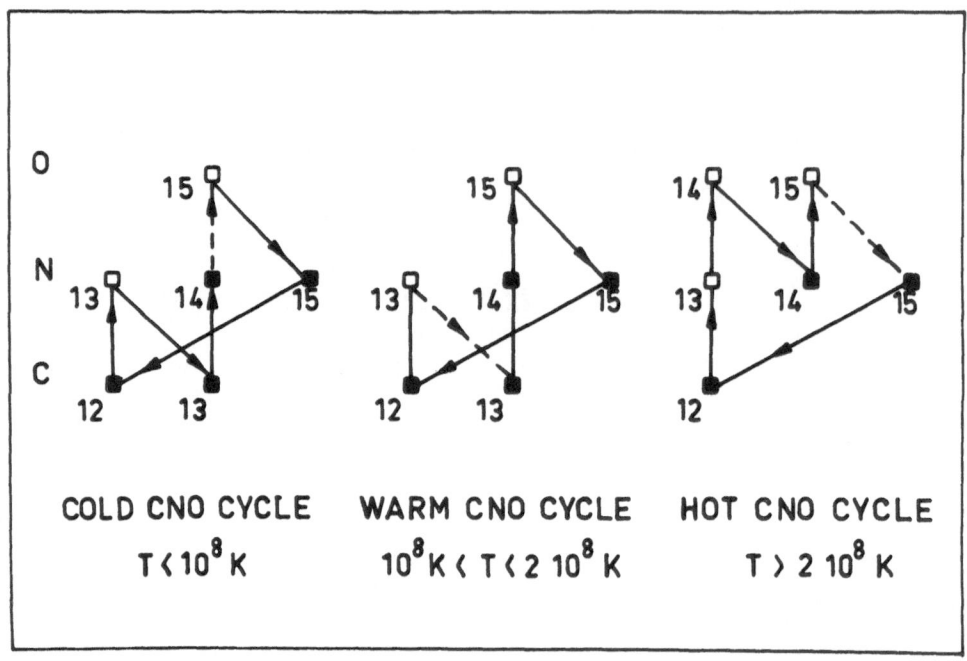

Figure 1 - Schematic networks for the cold (T < 10⁸ K), warm (10⁸ < T < 2 10⁸ K) and hot (T > 2 10⁸ K) CNO cycle. The bottle-neck (slowest reaction) is figured with dashed line for these three temperatures conditions.

The hot CNO cycle in hydrogen rich zones has been studied extensively, either at fixed temperature (Audouze et al, 1973 ; Audouze, 1973) or under the hydrodynamical time scale assumption (Arnould and Beelen, 1974). The results of these studies can be summarized as follows :

a) At low temperatures $T < 2 \times 10^8$ K, CNO transform mainly into ^{13}N (warm CNO cycle fig. 1) leading to substantial ^{13}C enrichment.

b) As T rises above 2×10^8 K the cycle bottlenecks are ^{14}O, ^{15}O, ^{17}F and ^{21}Na, progenitors of ^{14}N, ^{15}N, ^{17}O and ^{21}Ne. The ^{15}N/^{14}N ratio can be as large as 2 (while the solar system value is 3×10^{-3} and the equilibrium value at the end of the cold CNO cycle is $\sim 5 \ 10^{-5}$).

c) At $T \gtrsim 10^9$ K elements CNO transform into heavier elements and nuclei such as ^{19}F, ^{22}Ne, ^{24}Mg can be copiously produced.

Extensions of the reaction network to A < 11 elements by Arnould and Nogaard (1975) and Toussaint (1975) and to the range 25 < A < 35 by Toussaint (1975) and Chièze et al (1977) indicate that in similar conditions, elements such as ^{3}He, ^{7}Li and ^{11}B on one hand, and ^{25}Mg, ^{29}Si and ^{33}S on the other hand, might be easily produced.

To conclude this summary, astrophysical objects where the hot CNO-Ne cycle and its extensions can take place appear to be promising sites for the production of many relatively rare nuclear species.

III - HOT CNO NUCLEOSYNTHESIS WITH NOVA PROFILES

Together with super massive stars* (see e.g. Audouze and Fricke, 1973), novae are likely objects for which hot CNO cycle calculations are relevant. What is reported here is a summary of some results to be displayed elsewhere (Lazareff et al, 1977).

A network of 28 nuclei linked by more than 50 reactions has been run using a sample of various temperature-density profiles computed by Sparks, Starrfield and Truran (model 1 is still unpublished and model 2 is very similar to model 7 of Starrfiel et al, 1974).

The results of the nucleosynthesis depend mainly on the temperature, the initial composition and the freezeout timescale . The three temperature profiles (figured respectively in fig. 1, 2 and 3) selected here represent rather different types of nucleosynthesis. They illustrate respectively the warm (model 1), hot (model 2) and extreme (model 3) CNO cycle. In each case, the temperature rise is due to the thermonuclear runaway. When the degeneracy is raised, expansion and cooling occur. The conditions (temperature, density, composition) used here are those of the deepest zone of the model of Sparks et al. Of course one should take into account the subsequent mixing when one wants to estimate final enhancements.

Tables 1, 2 and 3 list respectively the final abundance enhancements achieved in models 1, 2 and 3.

The results of model 1 are a signature of the warm CNO cycle : the slowest reaction is $^{13}N(\beta^+)^{13}C$ therefore the biggest enhancement is that of ^{13}C ($\sim$ 1000) while ^{14}N, ^{15}N and ^{17}O are also but more moderatly enhanced ($\sim$ 50). In model 2, ^{15}N and ^{17}O are largely enhanced ($\sim$ 2000-5000) while ^{13}C and ^{21}Ne are moderatly enhanced. This is due to the relative slowness of $^{15}O(\beta^+)^{15}N$ and $^{17}F(\beta^+)^{17}O$. In model 3, the results depend strongly on the initial composition. When nucleosynthesis occurs in material very much enriched in ^{12}C (case a) there are not enough protons and alpha particles to burn it entirely. The dominant elements are C, N and O (^{12}C, ^{14}N, ^{15}N and ^{16}O). Although the temperature is large the reaction network cannot proceed up to higher mass. That is not the case when there are more protons and alpha particle available the

* If these still theoretical objects really exist !

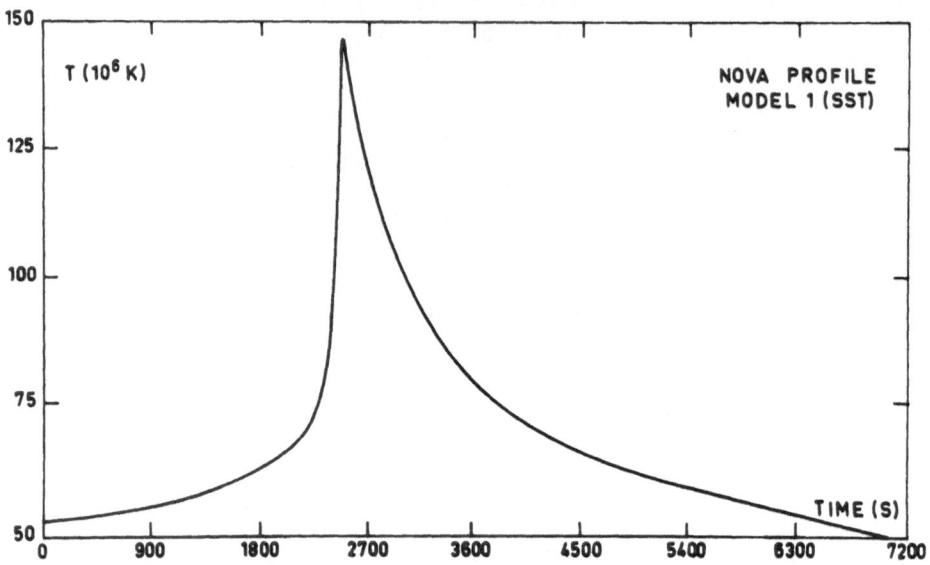

Figure 2 - Nova temperature profile.

TABLE 1

Nova Profile - Model 1   (Starrfield et al., unpublished)

Astrophysical Characteristics

$T_{max} = 1.47 \times 10^8 K - \langle\rho\rangle = 2.5 \times 10^3 g \ cm^{-3} - M_{env} = 2 \times 10^{29} \ g$

$- M_{ej} = 7 \times 10^{28} \ g$

Initial Abundances

$X = 0.37 \quad Y = 0.22 \quad Z(^{12}C) = 0.50$, other elements 0.5 solar

Final Enhancements (relative to solar)

$X = 0.25 \quad Y = 0.34$

$^{12}C = 24$	$^{16}O = 1$
$^{13}C = 1100$	$^{17}O = 85$
$^{14}N = 36$	$^{18}O \quad 4.4 \ (10^{-3})$
$^{15}N = 43$	the other elements are solar

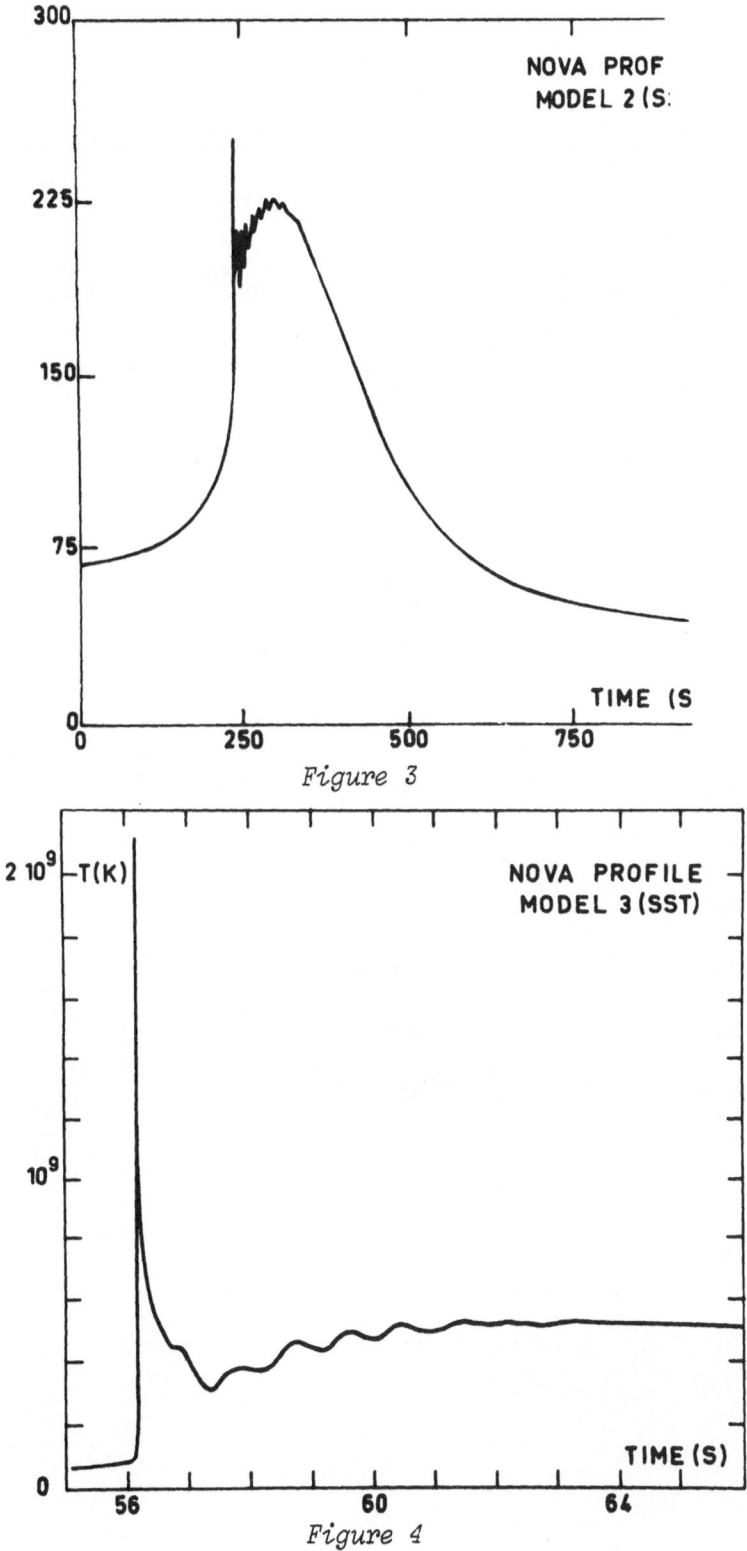

Figure 3

Figure 4

TABLE 2

Nova Profile - Modele 2   (Starrfield et al., unpublished)

Astrophysical Characteristics

T_{max} = 2.52 x 10^8 K - <ρ>= 1.4 x 10^4 g cm^{-3} - M_{env} = 2.5 x 10^{30} g

$- M_{ej}$ = 2.2 x 10^{29} g

Initial abundances

solar but $Z(^{12}C) = Z(^{16}O) = 0.04$

Final Enhancements

X = 0.54	Y = 0.38
^{12}C = 5.3	^{19}F = 0.12
^{13}C = 200	^{20}Ne= 1
^{14}N = 20	^{21}Ne= 200
^{15}N = 5400	^{22}Ne= 5
^{16}O = 4(-3)	^{23}Na= 4(-4)
^{17}O = 2200	^{24}Mg= 2
^{18}O = 8(-2)	^{25}Mg = 2

TABLE 3

Nova Profile – Model 3   Starrfield et al., 1975

Astrophysical Characteristics

$$T_{max} = 2.12 \times 10^9 \text{ K} - <\rho> = 1.6 \times 10^5 \text{ g cm}^{-3} - M_{env} = 3.2 \; 10^{30} \text{ g}$$

$$- M_{ej} = 8 \times 10^{29} \text{ g}$$

Initial Abundances

a   X, Y, Z (but not ^{12}C) = 0.1 x solar   Z(^{12}C) = 0.9

b   X, Y, Z (but not ^{12}C) = 0.2 x solar   Z(^{12}C) = 0.45

Final Enhancements

$$X = Y = 0$$

	a	b		a	b
^{12}C	120	2.4(-4)	^{19}F	2(-2)	4.2(+4)
^{13}C	40	4.1(-2)	^{20}Ne	2	0.50
^{14}N	410	3	^{21}Ne	3	8.7(+3)
^{15}N	2500	3.1(+4)	^{22}Ne	15	200
^{16}O	9	17	^{23}Na	.40	620
^{17}O	3(-6)	900	^{24}Mg	1.3	600
^{18}O	1.5(-2)	22	^{25}Mg	1	

reactions can proceed up to the higher mass (as shown already by Audouze et al, 1973, calculations for T_9 = 1) ^{17}O and ^{19}F are particularly enhanced. This is due to the effectiveness of $^{14}O(\alpha,p)^{17}F$ and $^{15}O(\alpha,\gamma)^{19}Ne$.

It is interesting to note that in this last model, at peak temperature, the dominant isotopes are ^{18}Ne and ^{18}O produced by $^{16}O(p,\gamma)$ ^{17}F (p,γ) ^{18}Ne, but soon destroyed by $^{18}Ne(^+\beta)$ ^{18}O (p,α) ^{15}N. As noticed by many workers, especially Nogaard (1977), and contrary to Cowan and Rose (1975), ^{18}Ne must not be forgotten in a proper CNO network.

To summarize, elements whose enhancement relative to solar abundance is the largest i.e. those for which the nova might be considered as a source are in model 1, ^{13}C ; in model 2, ^{15}N and ^{17}O, in model 3a, ^{15}N and in model 3b, ^{15}N, ^{19}F and ^{21}Ne.

A strong word of caution must be given about the large ^{15}N enhancements achieved in these calculations for two reasons :

First, the reaction $^{15}N(p,\alpha)^{12}C$ is still working at the base of the envelope (where the computations are made) when the outer layers are ejected. May be more important is the fact that the rate of that reaction is, at present, very uncertain (see Fowler et al, 1975). This must be recalled before drawing definitive conclusions on the ^{15}N synthesis.

Finally, in section II, the possibility of ^7Li production by $^3He(\alpha,\gamma)^7Be$ in similar temperature condition was brought up. Very recent calculations of Truran et al (1977) show that with profiles similar to those studied here, final abundances of ^7Li as large as $2 \sim 3\ 10^{-6}$ by mass can be achieved. In the case of this element, the problem is not so much to find a source for it as to choose among too many possible mechanisms (cosmological, red giants, novae, supernovae, supermassive stars, low energy cosmic rays ...).

IV - EXTREME NOVAE AND P PROCESS NUCLEOSYNTHESIS

The most extreme model used here (from Starrfield et al, 1975) has been designed to check an hypothesis of Clayton and Hoyle (1974) according to which novae could be a promising site for neutron production and therefore for s and also r process. Contrary to what is suggested in Starrfield et al (1975), in our calculations the number of neutrons obtained does not seem to be sufficient to induce such neutron capture nucleosynthesis.

However if these extreme models are not discarded on other astrophysical grounds, they appear to be promising sites of p process nucleosynthesis. P process elements are stable and have proton-rich nuclei which are bypassed when the neutron capture takes place. While their abundance curve is parallel to that of s and r elements these elements are rarer than s and r elements ($\sim$ 1% of their abundance) but their nucleosynthesis is still controversial matter. Among the different possibilities (such as spallation reactions or weak interactions) it has been suggested for instance by Truran and Cameron (1972) that they can be produced by explosive nucleosynthesis induced by protons on s or r elements or by (γ,n) reactions occuring at temperatures higher than 10^9 K. By following these arguments Audouze and Truran (1975) have recently explored the possibility of producing these isotopes by explosive nucleosynthesis induced in hydrogen rich zones. They have built seven different networks of about 200-300 nuclei each undergoing nuclear reactions induced by protons, alpha particles, and neutrons such as their reverse. Their calculations show that for temperatures $\sim$ 2 10^9 K and densities of $\sim$ 10^4 g cm^{-3} with characteritic free fall time scale of $449/\sqrt{\rho}$ i.e. $\sim$ 10 seconds significant and comparable enhancements of about 50-500 is obtained for 24 p process nuclei (on a total of 36 nuclei) i.e. 2/3 of them spread over the whole mass range. The interesting p process enhancements are indeed achieved after times as short as 10^{-3} 10^{-2} s comparable to the duration of the peak temperature of model 3. One must note that these calculations have been performed with extremely rough reaction rates and a very schematic treatment of the beta decay rates. Nevertheless it is the first and the only calculation (to our knowledge) which fits the p process nuclei over the whole mass range with a limited number of parameters. For instance, Woosley and Howard (1976) are advertizing a p process nucleosynthesis in oxygen burning zone which needs the tuning and the mixing of many zones having undergone different temperature-density conditions.

Therefore although the extreme nova model is far to be proved to really occur, it would be worthwhile to examine the possibility of such models since they appear to work so well for the p process nucleosynthesis.

V - NOVAE AND CHEMICAL EVOLUTION OF GALAXIES

With all the reservations made in section III, novae seem to be likely candidates to produce several rare isotopes and especially ^{15}N. For other nuclei such as ^{13}C for instance, other more " classical " sources such as red giants appear more promising. Up to very recently it was difficult to obtain information on the isotopic composition elsewhere than in the solar system. Hopefully many new determinations have become available after the study of various molecular clouds within the Galaxy. In this respect a group of the Bell Lab. has determined the variation of the $H^{12}C^{15}N/H^{13}C^{14}N$ ratio in various locations of the Galaxy (Linke et al, 1976).Their determination shows that this ratio is about two times lower in the solar neighborhood than in the solar system (0.18 instead of 0.33) but at least 10 times lower than the solar system value in the galactic center (< .026,1σ upper limit).

This interesting variation has been recently considered in terms of galactic evolution models by Audouze et al (1976). These authors note first that this strong decrease might not be due to a decrease of ^{15}N in the galactic center but presumably to a strong increase of ^{14}N correlated with the increase of ^{13}C. They have then performed a few calculations based on crude evolution models simulating the solar neighborhood and the galactic center (along the techniques described in Vigroux et al, 1976). They consider specifically two cases according to whether ^{15}N is produced in low mass or high mass stars. The conclusion is the following :

If ^{15}N is produced in novae with low mass progenitors, the predicted value of ^{12}C^{15}N/^{13}C^{14}N at the galactic center is twice the 1σ upper limit of Linke et al.(1976). This hypothesis is therefore only marginally acceptable. However although the novae are distributed like low mass stars, from the calculations showed in section III one cannot exclude the possibility that only a fraction of novae are active in the ^{15}N production. On the other hand, if ^{15}N is produced in objects with high mass progenitors (like supernovae), it is quite easy to reproduce the observed variation of the ^{12}C^{15}N/^{13}C^{14}N ratio.

More work, both observational and theoretical, is needed to establish or rule out the role of novae in the ^{15}N production on a galactic scale.

We thank Drs Warren Sparks, Sumner Starrfield and James W. Truran who have made their nova models available to us.

REFERENCES

Arnould, M. and Beelen, W., 1974, Astron. and Ap., 33, 215.
Arnould, M. and Norgaard, H., 1975, Astron. and Ap., 42, 55.
Audouze, J., 1973, in Explosive Nucleosynthesis, Schramm and
     Arnett ed. (Austin : U. of Texas Press).
Audouze, J. and Fricke, K.J., 1973, Ap.J., 186, 239.
Audouze, J., Lequeux, J., Rocca-Volmerange, B. and Vigroux, L.,
     1977 in CNO Isotopes in Astrophysics (J. Audouze ed., Reidel
     Publishing Co.).
Audouze, J. and Truran, J.W., 1975, Ap. J., 202, 204.
Audouze, J., Truran, J.W. and Zimmerman, B.A., 1973, Ap. J.,
     184, 493.
Chièze, J.P., Audouze, J. and Toussaint, J., 1977, in preparation.

Clayton, D.D. and Hoyle, F., 1976, Ap.J., 203, 490.

Cowan, J.J. and Rose, W.K., 1975, Ap.J., 201, L45.

Fowler, W.A., Caughlan, C.R. and Zimmerman, B.A., 1975, Ann. Rev. Astron. Ap., 13, 69.

Lazareff, B., Audouze, J., Starrfield, S. and Truran, J.W., 1977, in preparation.

Linke, R.A., Goldsmith, P.F., Wannier, P.G., Wilson, R.W. and Penzias, A.A., 1976, Ap.J., (submitted to publication).

Nogaard, H., 1977, in preparation.

Sparks, W.M., Starrfield, S. and Truran, J.W., this conference.

Starrfield, S., Sparks, W.M. and Truran, J.W., 1974, Ap.J. Suppl., 28, 247.

Starrfield, S., Truran, J.W. and Sparks, W.M., 1975, Ap.J., 198, L 113.

Toussaint, J., 1975, Thèse de 3e Cycle, Université Paris XI, unpublished.

Truran, J.W., Arnould, M. and Starrfield, S., 1977, in preparation.

Truran, J.W. and Cameron, A.G.W., 1972, Ap.J., 171, 89.

Vigroux, L., Audouze, J. and Lequeux, J., 1976, Astron. Astrophys. 52, 1.

OPTICALLY THICK WINDS IN CLASSICAL NOVAE AND SYMBIOTIC STARS

G.T. Bath

Department of Astrophysics, South Parks Road, Oxford.

ABSTRACT. The conditions in the photospheres of novae envelopes and in the hot components of symbiotic stars are discussed in terms of an outflowing wind which is optically thick. At outburst maximum in novae the mass loss rates are 10^{22} gm s^{-1}, which corresponds to equipartition between the energy flux of the gas and the radiation field. Radiation pressure in the continuum is the dominant accelerating force acting on the gas.

The decline of the optical light curve is produced by a fall in the mass loss rate. This generates a fall in the wind density, a decrease in photospheric radius, and an increase in effective temperature, as first pointed out by Grotrian (1937). Combined with a constant luminosity source (Gallagher and Code 1974), the resulting temperatures lead to the optical light being radiated progressively further into the Rayleigh-Jeans region of the continuum. The optical luminosity falls as a result of the increasing flux radiated in the unobserved UV. The conditions in the wind "photosphere" are shown to be approximately the same at the same decline stage and thus to account for the observed spectral appearance/decline stage correlation (McLaughlin 1960).

Similar processes seem to occur in the quasi-periodic outbursts of the symbiotic star Z And. It is suggested that in this case they are due to variable mass transfer by the giant companion at rates close to the critical Eddington limit rate.

REFERENCES

Gallagher, J.S. & Code, A.D., 1974, Astrophys.J. 189, 303.
Grotrian, W., 1937, Z.Astrophysics, 13, 215.
McLaughlin, D.B., 1960, Stars and Stellar Systems, VI.

M. Friedjung (ed.), Novae and Related Stars, 217. All Rights Reserved.
Copyright © 1977 by D. Reidel Publishing Company, Dordrecht, Holland.

Thermal Runaway near the Surface of Accreting White Dwarfs

Kyoji Nariai

Tokyo Astronomical Observatory, Mitaka, Tokyo 181

Ken-ichi Nomoto

Department of Astronomy, Faculty of Sciences, the University of Tokyo, Tokyo 113

and Dai-ichiro Sugimoto

Department of Earth Sciences and Astronomy, College of General Education, the University of Tokyo, Tokyo 153

Giannone and Weigert (1967) calculated models of white dwarfs on which a hydrogen rich envelope is assumed to increase with time. They have shown that a flash occurs for $\dot{M}=10^{-9} M_\odot$/year in their sequence D.   We have studied the same process in order to find out the conditions with which an accreting white dwarf can assimilate a nova phenomenon. When the rate of accretion $\dot{M}$ is large, the hydrogen-rich envelope is heated almost adiabatically. In this case the hydrogen is ignited near the surface and the flash is not strong enough to produce a nova. When $\dot{M}$ is too small, there is enough time for the excess energy to escape from the envelope and the ignition temperature is not reached. Strong flash occurs between these two cases. Mass of the degenerate core is also an important factor. Several sequences of models were calculated. A carbon-oxygen white dwarf of 1.3 solar mass with $\dot{M}=10^{-10} M_\odot y^{-1}$ makes a strong and rapid flash. We may expect a nova phenomenon for this set of parameters. For a helium white dwarf of 0.4 solar mass with $\dot{M}=10^{-8}$ and $10^{-13} M_\odot y^{-1}$, however, the flash process is very slow.

Reference
Giannone, P. and Weigert, A. 1967, Zeists. f. Astrophys, <u>67</u>, 41.

THE SLOW NOVA

Warren M. Sparks

Goddard Space Flight Center

Sumner Starrfield

Arizona State University

James W. Truran

University of Illinois

This model differs from our previous models, which resemble fast novae, in that it is a $1.25M_\odot$ white dwarf with no CNO enhancement in the hydrogen envelope. In this case, a shock wave is not generated, but the luminosity wave is. Because of the smaller radius associated with the larger mass white dwarf, the rekindled hydrogen shell source is stronger than in our less massive models. In fact, the luminosity is near the Eddington limiting luminosity at the shell source. Because of the higher opacity in the outer regions of the envelope, this high luminosity causes the outer regions to be ejected. The calculated light curve is very similar to that of a slow nova. This mechanism is also likely to cause the continuous mass loss in the remnant envelopes of our fast novae models.

THE EVOLUTION OF A NOVA MODEL WITH A Z=.03 ENVELOPE FROM PRE-EXPLOSION TO EXTINCTION

D. PRIALNIK, M.M. SHARA, G. SHAVIV

DEPARTMENT OF PHYSICS AND ASTRONOMY
TEL-AVIV UNIVERSITY
RAMAT-AVIV, TEL-AVIV, ISRAEL

A model for slow nova explosions is presented. The model consists of a 0.8 $M_\odot$ C/O core and an envelope of 10^{-4} $M_\odot$ with solar composition. The envelope is assumed to have been accreted from a companion. The nuclear runaway produces luminosity close to the Eddington luminosity ; this ejects 95% of the envelope.

We find
I   CNO equilibrium burning on a timescale of $10^{5.5}$ seconds produces enough energy for mass ejection.

II   The rise in luminosity stops close to the Eddington limit and the outer envelope layers accelerate via the continuous action of radiation pressure.

III   The mass outflow has two phases : a gentle outflow at the beginning and then a rapid outflow.

IV   The nova's "shut-off" mechanism is the exhaustion of the envelope's mass. In this slow nova model it took about 200 days for 95% of the envelope to be ejected and to leave behind a hot white dwarf.

V   The isotope ratios C^{12}/C^{13}, N^{14}/N^{15}, and O^{16}/O^{17} are in good agreement with observations.

VI   The behaviour of $L_{BOL}(t)$ agrees- at all wavelengths observed - well with observations. Several additional consequences are discussed.

HYDROGEN-RICH ENVELOPE MASS- A VITAL NOVA PARAMETER

M.M. SHARA, D. PRIALNIK, G. SHAVIV

DEPARTMENT OF PHYSICS AND ASTRONOMY
TEL-AVIV UNIVERSITY
RAMAT-AVIV, TEL-AVIV
ISRAEL

The behaviour of hydrogen-rich envelopes (Z=.03) of various masses on the surface of a 0.8 M_θ C/O white dwarf has been studied. The results can be summarized as follows:

I   For envelope masses below $2.6 \times 10^{-5} M_\theta$ no thermonuclear runaway occurs; the white dwarf cools continuously.

II   Above $2.6 \times 10^{-5} M_\theta$ the model undergoes a thermonuclear runaway of increasing strength. Close to the lower limit all the hydrogen burns into helium on a timescale of $\sim$ 100 years, without any mass-loss. The emitted luminosity during this period is 2-$3 \times 10^4 L_\theta$ at an effective temperature of $\sim 5 \times 10^5$ °K and a radius of $\sim$ .1 R_θ.

III   At $10^{-4} M_\theta$ the envelope is almost completely ejected by continuous mass-loss during $\sim$ 6 months; only $\sim$ 3% of the hydrogen initially present is consumed. We identify this model with a slow nova.
Considerably more massive ($\gtrsim 10^{-3} M_\theta$) envelopes lead probably to giant configurations burning on timescales longer than those common to novae.

M. Friedjung (ed.), Novae and Related Stars, 221. All Rights Reserved.
Copyright © 1977 by D. Reidel Publishing Company, Dordrecht, Holland.

Chairman Summary

Giora Shaviv

Dept. of Physics and Astronomy
Tel-Aviv University
Ramat Aviv, Israel

The basic theoretical effort exposed in this session was directed at the explanation of the very gross features of the Nova phenomenon e.g. energy requirements, total mass ejected, velocity in steady state outflow etc. So far very little was done to explain the fine features e.g. the internal structure of the outmoving envelope, the effect of the non-spherical symmetry imposed by the existing secondary star etc. The reasons are apparently the desire of theoreticians to get fast results or difficulties or both. Yet, the situation may also be justified on grounds that it is first essential to understand the overall physics before working out small details and in this respect we can consider the fine details as perturbations to an otherwise general picture.

Let me address myself first to the basic physical model. Are the causes of the various phenomena observed e.g. slow nova, fast nova, dwarf nova etc. different or do they form a gradual sequence of the very same physical process? The solution can be partly obtained theoretically by parameters analysis. Thus, we have to vary the input data and check how far we can go and still obtain the basic phenomenon. This is contrary to the method in which the parameters are chosen so as to agree best with a particular observed case. The same question is directed to the observers: to what extent are the various properties continuous or form some distinct classes. In this respect it is essential for the theoretical models to explain not only the light curve of a given object but also the various relations between different objects found by the observers e.g. (Δm, t3) relation.

M. Friedjung (ed.), Novae and Related Stars, 223-224. All Rights Reserved.
Copyright © 1977 by D. Reidel Publishing Company, Dordrecht, Holland.

The next step in the theoretical models should include
deviations from spherical symmetry. There are many observations
that indicate the importance of axial symmetry and hence the
dynamical models should include rotation, magnetic fields and
the comparison.

The present theoretical models seem to rely on the "radiation piston" to eject the outer layers. The interaction of
radiation and matter close to the Eddington limit is not fully
understood and the possible fluidization of the radiation may
give rise to luminosities in excess of the Eddington limit,
and to non-spherical symmetric outflow - namely mass-radiation
separation. The possible effect of "photon bubbles" is also
obscure at the moment.

The theoretical models have so far ignored the question of
relatively high accretion rates. The problem does not exist
for very low accretion rates for which the falling matter has
ample time to adjust its entropy density to that of the outer
layers of the accreting White Dwarf. In the initial configuration of the models discussed in this session the outer layers
are assumed to be in thermodynamic equilibrium. This will not
be the case in moderate and high accretion rates. Various
instabilities may arise in the outer layers when there is no
"adjustment" between the falling matter and the outer layers.
For example, the rate of rotation may be different giving rise
to shear instability and mixing via non-radial modes.

PART VI

FUTURE WORK

CHAIRMAN'S SUMMARY OF FINAL DISCUSSION : SESSION "FUTURE WORK"

J. SMAK

Copernicus Astronomical Center, Warsaw, Poland.

The following participants presented remarks during the
discussion: G. T. Bath, A. Cowley, P. P. Eggleton, M. Friedjung, R. P. Kraft, M. Livio, E. R. Mustel, K. Nariai, L. Rosino, W. Seitter, G. Shaviv, J. Smak, W. M. Sparks, H.-C.
Thomas, and B. Warner. The major topics were:

Origin and evolution. The most promising evolutionary
scheme is the one which should include: initial type C evolution of a relatively massive binary, followed by a rapid
phase of mass loss and momentum loss (presumably when the
system becomes a common envelope binary), and producing a
short period binary with a massive white dwarf. The relatively
long life-time of such a system would be consistent with indirect estimates of ages of cataclysmic variables based on
their distribution and kinematics. The onset and/or increase
in the mass transfer rate from the secondary component would
then mark the begining of the eruptive phase. A detailed
discussion of this scheme is most urgently needed.

Basic properties. While there is now a commonly accepted
standard model of a cataclysmic binary, more observational
data are needed concerning individual masses, rates of mass
transfer, and - particularly - properties of the circumstellar
matter. On the theoretical side, much is to be expected from
detailed studies of the mass transfer and mass loss processes,
structure of the disks, and the role of accretion. Of crucial
point here is the problem of viscosity which is being commonly
incorporated into the disk models but whose nature and origin
are far from being explained. This problem is important also
in the X-ray binaries.

M. Friedjung (ed.), Novae and Related Stars, 227-228. All Rights Reserved.
Copyright © 1977 by D. Reidel Publishing Company, Dordrecht, Holland.

Outbursts of U Gem variables. Again more observations
(including those in the uv) are badly needed to provide a
better description of the outburst phenomena. The role of the
instabilities in the secondary component and of the postulated
instabilities in the disk should be clarified on both observational and theoretical grounds. While the sudden accretion
mechanism for outbursts and other nova-like variations is
getting more and more support, there are still phenomena -
such as the rapid oscillations at outbursts, or the supermaxima of VW Hyi - which are not yet well understood.

Expanding envelopes of novae. Models of expanding envelopes based on detailed photometric and spectroscopic description of outbursts can provide now the important observational
counterpart to the theoretical models of outbursts. It appears
that the basic picture includes a major mass ejection at the
time of visual maximum followed by an extended period of continuous ejection. The details, however, of the acceleration
process and the origin of the time-dependent velocity fields
in the expanding envelope are to be worked out. The abundance
determinations - in view of their fundamental importance for
the theory - should also continue.

Outbursts of novae. The thermonuclear runaway models have
succeeded remarkably well in reproducing the major features
of a typical nova outburst. In particular, the overabundance
of CNO required by these models finds a direct counterpart in
the spectroscopic studies. It has also been shown that with
normal abundances the same mechanism can produce a slow nova.
There is, however, the disturbing problems of the initial
(evolutionary) conditions... If the hydrogen envelope is to
grow in mass due to accretion then it is not clear whether
the initial, less violent hydrogen burning will ever permit
to accumulate enough hydrogen rich mass in the envelope to
produce a major outburst. The problem of recurrent nove
remains completely open.

ASTROPHYSICS AND SPACE SCIENCE LIBRARY

Edited by

J. E. Blamont, R. L. F. Boyd, L. Goldberg, C. de Jager, Z. Kopal, G. H. Ludwig, R. Lüst,
B. M. McCormac, H. E. Newell, L. I. Sedov, Z. Švestka, and W. de Graaff

1. C. de Jager (ed.), *The Solar Spectrum, Proceedings of the Symposium held at the University of Utrecht, 26–31 August, 1963.* 1965, XIV + 417 pp.
2. J. Ortner and H. Maseland (eds.), *Introduction to Solar Terrestrial Relations, Proceedings of the Summer School in Space Physics held in Alpbach, Austria, July 15–August 10, 1963 and Organized by the European Preparatory Commission for Space Research.* 1965, IX + 506 pp.
3. C. C. Chang and S. S. Huang (eds.), *Proceedings of the Plasma Space Science Symposium, held at the Catholic University of America, Washington, D.C., June 11–14, 1963.* 1965, IX + 377 pp.
4. Zdeněk Kopal, *An Introduction to the Study of the Moon.* 1966, XII + 464 pp.
5. B. M. McCormac (ed.), *Radiation Trapped in the Earth's Magnetic Field. Proceedings of the Advanced Study Institute, held at the Chr. Michelsen Institute, Bergen, Norway, August 16–September 3, 1965.* 1966, XII + 901 pp.
6. A. B. Underhill, *The Early Type Stars.* 1966, XII + 282 pp.
7. Jean Kovalevsky, *Introduction to Celestial Mechanics.* 1967, VIII + 427 pp.
8. Zdeněk Kopal and Constantine L. Goudas (eds.), *Measure of the Moon. Proceedings of the 2nd International Conference on Selenodesy and Lunar Topography, held in the University of Manchester, England, May 30–June 4, 1966.* 1967, XVIII + 479 pp.
9. J. G. Emming (ed.), *Electromagnetic Radiation in Space. Proceedings of the 3rd ESRO Summer School in Space Physics, held in Alpbach, Austria, from 19 July to 13 August, 1965.* 1968, VIII + 307 pp.
10. R. L. Carovillano, John, F. McClay, and Henry R. Radoski (eds.), *Physics of the Magnetosphere, Based upon the Proceedings of the Conference held at Boston College, June 19–28, 1967.* 1968, X + 686 pp.
11. Syun-Ichi Akasofu, *Polar and Magnetospheric Substorms.* 1968, XVIII + 280 pp.
12. Peter M. Millman (ed.), *Meteorite Research. Proceedings of a Symposium on Meteorite Research, held in Vienna, Austria, 7–13 August, 1968.* 1969, XV + 941 pp.
13. Margherita Hack (ed.), *Mass Loss from Stars. Proceedings of the 2nd Trieste Colloquium on Astrophysics, 12–17 September, 1968.* 1969, XII + 345 pp.
14. N. D'Angelo (ed.), *Low-Frequency Waves and Irregularities in the Ionosphere. Proceedings of the 2nd ESRIN-ESLAB Symposium, held in Frascati, Italy, 23–27 September, 1968.* 1969, VII + 218 pp.
15. G. A. Partel (ed.), *Space Engineering. Proceedings of the 2nd International Conference on Space Engineering, held at the Fondazione Giorgio Cini, Isola di San Giorgio, Venice, Italy, May 7–10, 1969.* 1970, XI + 728 pp.
16. S. Fred Singer (ed.), *Manned Laboratories in Space. Second International Orbital Laboratory Symposium.* 1969, XIII + 133 pp.
17. B. M. McCormac (ed.), *Particles and Fields in the Magnetosphere. Symposium Organized by the Summer Advanced Study Institute, held at the University of California, Santa Barbara, Calif., August 4–15, 1969.* 1970, XI + 450 pp.
18. Jean-Claude Pecker, *Experimental Astronomy.* 1970, X + 105 pp.
19. V. Manno and D. E. Page (eds.), *Intercorrelated Satellite Observations related to Solar Events. Proceedings of the 3rd ESLAB/ESRIN Symposium held in Noordwijk, The Netherlands, September 16–19, 1969.* 1970, XVI + 627 pp.
20. L. Mansinha, D. E. Smylie, and A. E. Beck, *Earthquake Displacement Fields and the Rotation of the Earth, A NATO Advanced Study Institute Conference Organized by the Department of Geophysics, University of Western Ontario, London, Canada, June 22–28, 1969.* 1970, XI + 308 pp.
21. Jean-Claude Pecker, *Space Observatories.* 1970, XI + 120 pp.
22. L. N. Mavridis (ed.), *Structure and Evolution of the Galaxy. Proceedings of the NATO Advanced Study Institute, held in Athens, September 8–19, 1969.* 1971, VII + 312 pp.
23. A. Muller (ed.), *The Magellanic Clouds. A European Southern Observatory Presentation: Principal Prospects, Current Observational and Theoretical Approaches, and Prospects for Future Research, Based on the Symposium on the Magellanic Clouds, held in Santiago de Chile, March 1969, on the Occasion of the Dedication of the European Southern Observatory.* 1971, XII + 189 pp.

24. B. M. McCormac (ed.), *The Radiating Atmosphere. Proceedings of a Symposium Organized by the Summer Advanced Study Institute, held at Queen's University, Kingston, Ontario, August 3–14, 1970.* 1971, XI + 455 pp.

25. G. Fiocco (ed.), *Mesospheric Models and Related Experiments. Proceedings of the 4th ESRIN-ESLAB Symposium, held at Frascati, Italy, July 6–10, 1970.* 1971, VIII + 298 pp.

26. I. Atanasijević, *Selected Exercises in Galactic Astronomy.* 1971, XII + 144 pp.

27. C. J. Macris (ed.), *Physics of the Solar Corona. Proceedings of the NATO Advanced Study Institute on Physics of the Solar Corona, held at Cavouri-Vouliagmeni, Athens, Greece, 6–17 September 1970.* 1971, XII + 345 pp.

28. F. Delobeau, *The Environment of the Earth.* 1971, IX + 113 pp.

29. E. R. Dyer (general ed.), *Solar-Terrestrial Physics/1970. Proceedings of the International Symposium on Solar-Terrestrial Physics, held in Leningrad, U.S.S.R., 12–19 May 1970.* 1972, VIII + 938 pp.

30. V. Manno and J. Ring (eds.), *Infrared Detection Techniques for Space Research. Proceedings of the 5th ESLAB-ESRIN Symposium, held in Noordwijk, The Netherlands, June 8–11, 1971.* 1972, XII + 344 pp.

31. M. Lecar (ed.), *Gravitational N-Body Problem. Proceedings of IAU Colloquium No. 10, held in Cambridge, England, August 12–15, 1970.* 1972, XI + 441 pp.

32. B. M. McCormac (ed.), *Earth's Magnetospheric Processes. Proceedings of a Symposium Organized by the Summer Advanced Study Institute and Ninth ESRO Summer School, held in Cortina, Italy, August 30–September 10, 1971.* 1972, VIII + 417 pp.

33. Antonin Rükl, *Maps of Lunar Hemispheres.* 1972, V + 24 pp.

34. V. Kourganoff, *Introduction to the Physics of Stellar Interiors.* 1973, XI + 115 pp.

35. B. M. McCormac (ed.), *Physics and Chemistry of Upper Atmospheres. Proceedings of a Symposium Organized by the Summer Advanced Study Institute, held at the University of Orléans, France, July 31–August 11, 1972.* 1973, VIII + 389 pp.

36. J. D. Fernie (ed.), *Variable Stars in Globular Clusters and in Related Systems. Proceedings of the IAU Colloquium No. 21, held at the University of Toronto, Toronto, Canada, August 29–31, 1972.* 1973, IX + 234 pp.

37. R. J. L. Grard (ed.), *Photon and Particle Interaction with Surfaces in Space. Proceedings of the 6th ESLAB Symposium, held at Noordwijk, The Netherlands, 26–29 September, 1972.* 1973, XV + 577 pp.

38. Werner Israel (ed.), *Relativity, Astrophysics and Cosmology. Proceedings of the Summer School, held 14–26 August, 1972, at the BANFF Centre, BANFF, Alberta, Canada.* 1973, IX + 323 pp.

39. B. D. Tapley and V. Szebehely (eds.), *Recent Advances in Dynamical Astronomy. Proceedings of the NATO Advanced Study Institute in Dynamical Astronomy, held in Cortina d'Ampezzo, Italy, August 9–12, 1972.* 1973, XIII + 468 pp.

40. A. G. W. Cameron (ed.), *Cosmochemistry. Proceedings of the Symposium on Cosmochemistry, held at the Smithsonian Astrophysical Observatory, Cambridge, Mass., August 14–16, 1972.* 1973, X + 173 pp.

41. M. Golay, *Introduction to Astronomical Photometry.* 1974, IX + 364 pp.

42. D. E. Page (ed.), *Correlated Interplanetary and Magnetospheric Observations. Proceedings of the 7th ESLAB Symposium, held at Saulgau, W. Germany, 22–25 May, 1973.* 1974, XIV + 662 pp.

43. Riccardo Giacconi and Herbert Gursky (eds.), *X-Ray Astronomy.* 1974, X + 450 pp.

44. B. M. McCormac (ed.), *Magnetospheric Physics. Proceedings of the Advanced Summer Institute, held in Sheffield, U.K., August 1973.* 1974, VII + 399 pp.

45. C. B. Cosmovici (ed.), *Supernovae and Supernova Remnants. Proceedings of the International Conference on Supernovae, held in Lecce, Italy, May 7–11, 1973.* 1974, XVII + 387 pp.

46. A. P. Mitra, *Ionospheric Effects of Solar Flares.* 1974, XI + 294 pp.

47. S.-I. Akasofu, *Physics of Magnetospheric Substorms.* 1977, XVIII + 599 pp.

48. H. Gursky and R. Ruffini (eds.), *Neutron Stars, Black Holes and Binary X-Ray Sources.* 1975, XII + 441 pp.

49. Z. Švestka and P. Simon (eds.), *Catalog of Solar Particle Events 1955–1969. Prepared under the Auspices of Working Group 2 of the Inter-Union Commission on Solar-Terrestrial Physics.* 1975, IX + 428 pp.

50. Zdeněk Kopal and Robert W. Carder, *Mapping of the Moon.* 1974, VIII + 237 pp.

51. B. M. McCormac (ed.), *Atmospheres of Earth and the Planets. Proceedings of the Summer Advanced Study Institute, held at the University of Liège, Belgium, July 29–August 8, 1974.* 1975, VII + 454 pp.

52. V. Formisano (ed.), *The Magnetospheres of the Earth and Jupiter. Proceedings of the Neil Brice Memorial Symposium, held in Frascati, May 28–June 1, 1974.* 1975, XI + 485 pp.

53. R. Grant Athay, *The Solar Chromosphere and Corona: Quiet Sun.* 1976, XI + 504 pp.

54. C. de Jager and H. Nieuwenhuijzen (eds.), *Image Processing Techniques in Astronomy. Proceedings of a Conference, held in Utrecht on March 25–27, 1975,* XI + 418 pp.

55. N. C. Wickramasinghe and D. J. Morgan (eds.), *Solid State Astrophysics. Proceedings of a Symposium, held at the University College, Cardiff, Wales, 9–12 July 1974.* 1976, XII + 314 pp.

56. John Meaburn, *Detection and Spectrometry of Faint Light.* 1976, IX + 270 pp.

57. K. Knott and B. Battrick (eds.), *The Scientific Satellite Programme during the International Magnetospheric Study. Proceedings of the 10th ESLAB Symposium, held at Vienna, Austria, 10–13 June 1975.* 1976, XV + 464 pp.

58. B. M. McCormac (ed.), *Magnetospheric Particles and Fields. Proceedings of the Summer Advanced Study School, held in Graz, Austria, August 4–15, 1975.* 1976, VII + 331 pp.

59. B. S. P. Shen and M. Merker (eds.), *Spallation Nuclear Reactions and Their Applications.* 1976, VIII + 235 pp.

60. Walter S. Fitch (ed.), *Multiple Periodic Variable Stars. Proceedings of the International Astronomical Union Colloquium No. 29, Held at Budapest, Hungary, 1–5 September 1975.* 1976, XIV + 348 pp.

61. J. J. Burger, A. Pedersen, and B. Battrick (eds.), *Atmospheric Physics from Spacelab. Proceedings of the 11th ESLAB Symposium, Organized by the Space Science Department of the European Space Agency, held at Frascati, Italy, 11–14 May 1976.* 1976, XX + 409 pp.

62. J. Derral Mulholland (ed.), *Scientific Applications of Lunar Laser Ranging. Proceedings of a Symposium held in Austin, Tex., U.S.A., 8–10 June, 1976.* 1977, XVII + 302 pp.